# Reflections on
# Richard Hartshorne's
# *The Nature of Geography*

# Reflections on
# Richard Hartshorne's

# *The Nature*
# *of Geography*

## Occasional Publications of the
## Association of American Geographers

Series Editor:
Anthony R. de Souza
*National Geographic Society*

Editors:
J. Nicholas Entrikin
*University of California, Los Angeles*
and
Stanley D. Brunn
*University of Kentucky*

# AAG

Library of Congress Catalog Number 89-083425

ISBN 0-89291-204-9

# Contents

# Series Editor's Preface

Reflections on Richard Hartshorne's *The Nature of Geography* is the first volume of a series of Occasional Publications of the Association of American Geographers. This series is based on the work of specialists; the task of the volumes' authors or editors is to decide what is significant and interesting in a particular domain. Taken together, the volumes offer an intelligent guide to the geographer's craft.

This edited volume reflects on Richard Hartshorne's *The Nature of Geography*, a searching and meticulous review of the historical development of geography. When *The Nature of Geography* was written some fifty years ago, geography was a subject grasping for discipline. Hartshorne synthesized the work of many scholars who had gone before—largely scholars from Germany, France, the British Isles, and the United States. His accomplishment was extraordinary. His book facilitated the study of national geographies as had never been done before, brought a mature European point of view to the geography of North America, and contributed to the internationalization of American geography.

*The Nature of Geography* provides all of us with a penetrating insight into the intellectual highways and byways of the field, pointing out the paths that have been trod and by whom and when. It is not an exhaustive synthesis; instead, it is an excursion through a mass of literature, sometimes little known, and an attempt to reveal the views that prevailed at various points in time. To be sure, the manuscript is dominated by the thinking of German geography, but Germany had long dominated the field. If other geographies developed after World War II, the legitimacy of Hartshorne's concern with a previous German geography was never in question; and if the direction of geography changed in the postwar era, the validity of the work published in 1939 still held. Hartshorne's book anticipated that geography was a subject in evolution. And so *The Nature of Geography* is a rigorous contribution of timeless value. It continues to provoke questions about the nature of the discipline, the place of geography in the history of science, and the position of geography in relation to other fields of scholarly endeavor. Despite criticisms of Hartshorne's massive work, some of which were answered in *Perspective on the Nature of Geography*, *The Nature of Geography* continues to provide us with an effective framework for understanding disciplined growth and for explaining that growth to fellow professionals.

Richard Hartshorne attended his first meeting of the Association of American Geographers in Cincinnati in 1923. Sixty-six years later he attended the 1989 meeting of the Association in Baltimore. Who else can claim such a span of attendance? And over all those years, Hartshorne has shared his thoughts with students and colleagues. *The Nature of Geography* is a part of that sharing. It is to this remarkable achievement that the first volume of the series is gratefully dedicated.

Anthony R. de Souza
*Bethesda, Maryland, August 1989*

# Editors' Preface

Fifty years ago the *Annals of the Association of American Geographer* published a two-part article written by Richard Hartshorne on geographic methodology. In the same year, the article was reprinted as a monograph entitled *The Nature of Geography*. It has since become one of the most influential works in American geography, especially during the middle decades of the twentieth century. For the last several decades, its arguments have been caricatured more often than they have been seriously engaged. The fiftieth anniversary of the publication of *The Nature* offered an opportunity for a reconsideration of its arguments in terms of recent developments in the field, especially the resurgence of interest in place and region. The shape of this reconsideration has evolved over several years, and its final form is quite different from the original conception. A brief history of the project may be useful for the reader to help understand its present format.

The idea of a fiftieth-anniversary, critical reading in light of current interests was proposed by Nick Entrikin to the former editor of the *Annals*, Susan Hanson. After a discussion with the Editorial Board of the *Annals* and the Publications Committee of the Association, it was decided that a special section of the *Annals* would be devoted to four or five short papers commenting on the contemporary significance of *The Nature*. It was also decided that all who would wish to contribute would be given an opportunity to submit a paper. A notice was placed in the AAG *Newsletter*, inviting participation. The initial five papers were a combination of both solicited and unsolicited manuscripts. The attempt was made initially to shape the discussion toward contemporary geography, and away from too great an emphasis on the well-trodden and littered battleground of regional geography versus spatial science, without, of course, ignoring this important debate in Anglo-American geography. No attempt was made to establish uniformity or conformity of interpretation among the essays. The views of the authors are often at odds with other contributors and with Hartshorne himself. The authors express interests in a variety of themes, including early twentieth-century German geographic thought, American geography at mid-century, key personalities in debates, social theory, marxism, place, and region.

Stan Brunn succeeded Susan Hanson as editor of the *Annals* in 1987 and took over the responsibilities of guiding the papers into print. The initial papers were extensively reviewed for *Annals* publication. Concerns about journal space and the possibility of publishing the papers as a separate volume led to a relatively late change in format. The *Occasional Publications Series* of the Association, under the editorship of Tony deSouza, had been initiated, and it was thought that the set of papers would constitute an appropriate inaugural volume of this series. This decision was made easier by the lifting of page restrictions to allow the inclusion of several papers on related themes that came late to the attention of the editors. Among these were papers from a separately planned but overlapping event concerning *The Nature* that David Ward arranged as a Presidential Plenary Session at the 1989 meeting of the Association in Baltimore.

Thus the project is in some ways an Association project involving several committees, editors, and elected representatives at various times in the planning process. The papers indirectly illustrate this point, in that their diverse viewpoints cover many of the wide-ranging epistemological, theoretical and thematic concerns evident within the discipline. This volume is an intentionally presentist contribution to the study of geographic thought that says as much about geography in the 1980s as it does about geography in the 1930s. The central unifying element (or, in Professor Hartshorne's words, "target") is *The Nature.*

It is our hope that this volume will be studied by those European and American geographers interested in and familiar with the past half-century of geographic thought, by undergraduate and graduate students in philosophy and methodology classes, and those seeking directions for geography as a discipline in the twenty-first century. We believe there is much in these essays that merits the attention of those interested in regionalism, landscapes, neo-Kantian influences, social theory, and twentieth-century European geography as well as the evolution of geography as a discipline.

Some of the many people who have contributed to this volume have already been noted, but we would like to add to that list. The editors would like to thank Vincent Berdoulay, Steve Daniels, Harm deBlij, George Kish, Richard Morrill, Jim Wheeler, Phil Wagner, Don Janelle, Rose Canon, and Sharon Kindall. We would also like to thank Richard Hartshorne for his active interest and participation in this project.

J. Nicholas Entrikin
Stanley D. Brunn

# Introduction:
# *The Nature*
# *of Geography*
# in Perspective

J. NICHOLAS ENTRIKIN

Department of Geography, University of California, Los Angeles, CA 90024

Fifty years after its appearance in the pages of two numbers of the *Annals of the Association of American Geographers*, Richard Hartshorne's *The Nature of Geography* (1939) remains a frequently cited work in Anglo-American geography. It has achieved the status of a "classic" in geographic thought and has become a symbol of a particular era and style of geographical research. Geography students are assigned parts of *The Nature* in their geographic thought courses, but references to it suggest that authors seek to evoke a complex of concerns that the book has come to symbolize rather than to reconsider and carefully analyze Hartshorne's densely textured arguments. Such a fate is common to many classic works in the human sciences, but it is nonetheless unfortunate. The arguments found in *The Nature* and those attributed to it as the symbol of a descriptive regional geography diverge at many points. The recognition of this difference is of potential importance to geographers and social scientists in the late twentieth century as the topic of areal variation re-emerges in the discourse of the human sciences.

The significance of Hartshorne's still-impressive work for contemporary geography is in its recognition and analysis of the logical problems associated with the objective study of the specificity of place and region. Through his analysis of the German methodological literature and in his own arguments, Hartshorne sought to resolve a fundamental tension in the science of geography between its spatial perspective, that "sees together" the heterogeneous phenomena that constitute place and region, and the logical requirements of scientific concept formation. He valued both scientific rationality and the geographer's concern with the specificity of place and region, and sought to balance these seemingly contradictory goals.

Hartshorne's *The Nature* and later the *Perspective on the Nature of Geography* (1959) are the two most exhaustive analyses written in the English language concerning the logical problems faced in accommodating this naive sense of the differences among places with the objective, decentered view of the scientist. Geographers continue to express these concerns, which would seemingly imply that *The Nature* would be one of the standard texts in modern geographical epistemology. My impression is, however, that most geographers do not view *The Nature* in this way, but rather view it primarily as a window for looking into a geographic past. This attitude is somewhat surprising given the revival of interest in the study of areal variation during the 1980s.

I shall suggest that *The Nature* offers modern geographers insights into the issues associated with the logic of the study of the particular, and that such insights are of use in the renewed concern with specific place, region and landscape. These insights are limited, however, by the relatively constricted meaning of place and region that is found in *The Nature,* a limitation that was quite consistent with the concerns of its author and more generally with the mores of early twentieth-century American geography. One of the epistemological concerns that guided Hartshorne was to establish the objectivity of chorological study. A consequence of this emphasis in his work was the relative neglect of some of the more subjective aspects of our experience of place that make it such a rich concept for understanding modern life. This emphasis on objectivity has contributed to the relative neglect of *The Nature* in late twentieth-century discussions of place.

Hartshorne recognized the impossibility of characterizing the specificity of place and region from the decentered view of the theoretical scientist. He was also aware of the inherent subjectivity associated with such studies. He did not, however, wish to give up the goal of scientific objectivity, and thus he limited his arguments to the facts of place. To many modern geographers, this limitation seems severe. The chorological concept of place and region covers only a small part of the wide band that constitutes the meaning of place. Or, to use the geological metaphor that Denis Cosgrove (1985) applies to the morphological view of landscape, chorologists gain access only to the surface level of meaning.

The chorological viewpoint presented by Hartshorne does not incorporate those aspects of place and region associated with our awareness of being subjects situated in the world. We encounter place as a condition of our experience. As agents we are always "in place," in the same way that we are always "in culture" (Richardson 1984). Our relation to place is a part of our individual and collective identities. As Tuan (1974, 213) states: "Place is not only a fact to be explained in the broader frame of space, but it is also a reality to be clarified and understood from the perspectives of the people who have given it meaning." Hartshorne addressed the first half of this dualism and bracketed the second as outside of the realm of a science of chorology (Sack, this volume). For example, he recognized the place of aesthetic geography in Hettner's work, but was careful to point out that it had a role in geography only to the extent that it could be objective (Hartshorne 1939; Butzer, this volume). Its potential for objective analysis distinguished it from what Hartshorne deemed to be the unacceptable subjectivity associated with the idea of geography as art.

Hartshorne was well aware of the ease with which geographic concerns cross over the various discourses associated with everyday life, art, and science. In seeking to carve out an intellectual space for a science of regions, however, he accentuated the differences between these various modes rather than their points of contact. In contemporary discussions of place, region and landscape, the emphasis is given to these points of contact and to the areas of overlap among discourses. Hartshorne's arguments appear to be somewhat restrictive to the student of place and region who views the meaning of these concepts as including both an objective and a subjective reality (Sack, this volume).

By raising the issue of Hartshorne's emphasis on the relatively objective meanings of place and region, I do not mean to belittle the contribution of *The Nature*, which has been enormous, but rather to put in better focus a part of this contribution. For example, Hartshorne's arguments have direct relevance for the so-called "new," "theoretically informed" regional geography (Agnew, this volume). One of the central methodological themes of *The Nature* concerned the question of the criteria of significance and selection in the study of place and region. The intellectual "break" between the meta-theoretical introductions to recent studies of place and region and narrative-like syntheses that follow these introductions suggests that the problem of establishing appropriate criteria of significance remains an issue despite the frequent reference to theory.

# Current Views of *The Nature*

The relevance of Hartshorne's arguments for contemporary geography is not readily apparent in the geographical literature. A review of this literature gives the impression of two prevailing, and somewhat distinct, readings of *The Nature*, one as a text in the history of geographic thought and the other as a methodological treatise. The tendency to distinguish the historical from the logical offers an insight into the distance between the spirit that informed *The Nature* versus that which informs contemporary geography. The goal of discovering the "core" of geography through an historical and logical analysis of what professional geographers have said about their practices seems suspect to historians of geographers in the late twentieth century (for example, Stoddart 1986). For this reason discussions of *The Nature* as a text in the history of geographic thought tend to characterize it as an artifact, as opposed to an exemplar. Our textbooks continue to rely upon Hartshorne's interpretations of the works of nineteenth-century German geographers, but historians of geography tend not to imitate his style of analysis.

The logical arguments of *The Nature* fare only slightly better. Hartshorne's arguments are perceived as having provided the model for a currently out-of-favor, atheoretical and positivistic conception of geography as areal differentiation. Some contend that the philosophical differences between Hartshorne and his critic Schaefer were more apparent than real, and that Hartshorne's arguments were a precursor to a positivistic spatial science (for example, Gregory

1978; Guelke 1978; Entrikin 1981; Lukermann, this volume). In general, however, the references are to *The Nature* as an artifact of a geographic past. Hartshorne's arguments concerning areal variation are seen as descriptive, and hence unscientific by the spatial analyst, as too atheoretical by the neo-Marxist, and as too positivistic by the humanist. Postmodernists have favored the study of areal variation and have outlined their vision of a "polyvocal" geography, but Hartshorne's has not been one of the voices that has been recalled from the geographic past (Dear 1988; Soja 1989).

These diverse criticisms offer insight into a notable silence in the contemporary literature concerning *The Nature.* The return of areal variation in the study of specific places, regions and landscapes to the research agenda of contemporary geography has not appeared to have generated a concern for a reexamination of *The Nature* (Agnew, this volume). References to *The Nature* are part of the recent geographical literature, but they rarely reflect an active engagement with the arguments expressed in that work. Consideration of both the presence and the absence of reference to *The Nature* raises the question of why Hartshorne's work is not seen as a more significant and a more positive part of modern geographical discourse.

Many reasons could be offered to help answer this question (Smith; Agnew, this volume). I shall limit myself here to two related responses. The first concerns certain anachronistic qualities of Hartshorne's argument, associated with his attempt to derive a view of the field through what he described in his subtitle as "A Critical Survey of Current Thought in Light of the Past." Hartshorne characterizes his work as in part an empirical investigation of what geographers have said about their field.

The second concerns the rather single-dimensional critical interpretation of the logical arguments of *The Nature* that equated chorology with the study of the unique. This interpretation has dominated recent discussions of the study of place and region. These two issues are related in that Hartshorne's sense of his project emphasized accurate representation of what I shall call the "classics" of German geography (Alexander 1989; Elkins, this volume). When challenged, he responded in terms of historical and textual accuracy. He appeared to view the primary grounding of his argument as being empirical rather than logical. This emphasis was not unusual in the context of debate in American geography during the first half of the twentieth century. American science in the early twentieth century looked toward German scholarship as a model, and geography was no exception to this generalization (Elkins; Butzer; Lukermann; Smith; this volume). For example, an important part of the authority of Sauer's methodological articles was no doubt tied to his ability to draw support from the German literature.

Such authority is never unchallenged, however, and always changing. A clear illustration of this fact is evident in the spatial-analytic attack on Hartshorne's work. Schaefer's (1953) criticism was based on a reading of some of the same sources used by Hartshorne. The response by Hartshorne (1955; Martin, this volume) was the empirical one of demonstrating how Schaefer had misrepresented the classics. The grounds had shifted, however, in that the supporters of Schaefer were less concerned with the accuracy of Schaefer's translations than they were with the perceived power and potential of the model of geography that he offered. Hartshorne's concerns over scholarship fell on relatively deaf ears. Authority appeared now to be embedded in the philosophy of science and, more specifically, in the arguments of logical empiricism. A new set of classics were added to the old, but authority resided outside of geography, as seems to be the case today.

Hartshorne's empirical emphasis left the more epistemological concerns of *The Nature* only partially expressed and thus opened to reinterpretation by the more philosophically grounded arguments of the spatial analysts. The translation of several of the neo-Kantian themes found in *The Nature* into the language of logical empiricism reduced significantly their meaning and coherence (Entrikin 1981). This tendency is most clearly illustrated in the translation of the idiographic into the study of unique objects.

## Anachronistic Character

One of the most notable characteristics of *The Nature* is its inwardness. Its emphasis on what professional geographers have said about their field contrasts with the more recent tendency to place high value on what scholars outside of geography have said about place, space and landscape. The inward-looking quality of geography in the mid-twentieth century contrasts significantly with the outward-looking geography of the present (Smith; Sack, this volume).

Contemporary geographers continue to consider their past, but generally not with the hope of uncovering a geographic "core." Our current concern with the history of geographical thought has not had the effect of building a consensus about the practices of the discipline, but rather has seemed to further divide an already fragmented field. Indeed, the hope that consensus could be achieved by the study of the past seems somewhat naive to the modern viewpoint. Ours is a relativistic age.

Our late twentieth-century doubts about progress and consensus are expressed in views about how the history of our field should be written. Historians of geography endorse what they refer to as a contextual model, in which geographical ideas are seen in the wide view of the social events and intellectual currents of their day (Berdoulay 1981). They contrast this model with that of a normative reconstruction model in which a relatively linear progress of ideas is traced through time. The contextual model has been used to express a more contingent character of the history of geographic ideas. Despite Hartshorne's defense of the empirical quality of his work, it is now generally viewed as a normative reconstruction of geographic thought (for example, Stoddart 1986; Johnston 1979; Gregory 1978). It is also an internalist view, a view of geography from the perspective of professional geographers. Both genres of the history of geographic thought continue to have practitioners, but they are somewhat at odds with the contextualist and historiographic spirit of modern historians of geography.

The arguments in *The Nature* reflect a time in which the battle was being waged over the core of geography, in marked opposition to the 1980s in which battles seem to be waged in skirmishes on the periphery. This lack of centrality has been vividly described by David Stoddart (1986, 55) in his characterization of the revolutionary tendencies in modern geography, when, "at the call of 'Charge!,' the geographical horsemen thunder past at an increasingly frenetic rate, only shortly to learn that the real onslaught needs to be made elsewhere." Indeed, it is not clear that the core-periphery analogy has any meaning in contemporary geography. The dense textual criticism, which characterized Hartshorne's interpretations of what he took to be the central methodological studies in geography, is now more likely to be found in the modern geographer's interpretations of philosophical and social-theoretical texts.

The possibility of establishing a consensus concerning the identification of a geographical core would seem much less in the late twentieth century as opposed to the middle of the century. We must be careful, however, not to overstate the degree of consensus in past periods. Sauer's (1941) criticism of *The Nature* in his 1940 Presidential Address to the AAG illustrates the strong differences that existed. These differences in viewpoint did not appear, however, to place in question the belief that a core existed and that it could be uncovered. The underlying belief in the possibility of consensus is well illustrated in the understated criticism of a British reviewer, J. L. Myres (1940, 399), who summarized his view of Hartshorne's argument by stating that:

> All this does not go far beyond what British geographers are practising; but it is useful to know what the transatlantic difficulties are, and encouraging to find that a "logical" solution of them supports our common sense.

Geography in the late twentieth century appears too diverse to allow for the coming together of logic and commonsense, or epistemology and practice.

## Problems of Interpretation

The logical arguments that support the chorological position are not well understood in contemporary geography. The limited role of *The Nature* in recent discussions of place-studies has been in part a function of the persistence of the spatial-analytic interpretation of chorology. This interpretation simplifies the Hartshornian argument concerning the study of the specific region by reducing its epistemological dimensionality.

Hartshorne has consistently maintained that the issue of the idiographic and the nomothetic and the related issue of exceptionalism, a term applied to Hartshorne's arguments by Schaefer (1953), have been misunderstood by his critics. The basis of the misreading, according to Hartshorne, is the failure to recognize the argument that geography is both a nomothetic and an idiographic discipline.

For Hartshorne, the arguments of the spatial analysts, as well as more generally those of theoretical geographers, do not contradict his central themes. He sees them as compatible with his interpretation of the field, except for their failure to recognize the importance of the idiographic concern with specific place and region.

Hartshorne's reaction to his critics has often been ignored. The debates between regional geographers and spatial analysts continue to be described in terms of the study of the unique versus the search for lawful generalization, description versus theory, or proto-science versus science. It remains a curious fact of the recent history of geographic thought that the arguments of those opposed to the conception of geography as a positivist spatial science have somewhat uncritically accepted the spatial-analytic interpretation of *The Nature*.

As a step toward understanding the modern relevance of *The Nature*, it is necessary to reconstruct the interpretation so that it derives more from the work itself, rather than from critical interpretation of the work. Such a reconstruction begins with a better understanding of the idiographic and the nomothetic, and its association with the geographer's concern to develop a scientific logic for the study of the particular. To help characterize this concern, I find it useful to quote from John Leighly's 1937 *Annals* article, "Some Comments on Contemporary Geographic Method." In this work, Leighly highlights one of the central logical puzzles facing a scientific chorological perspective. The arguments of Leighly were not only an important part of the intellectual context of Hartshorne's work, but also presaged many of the issues that were raised during the 1960s and 1970s by critics, and that continue to be raised today. Leighly (1937, 128) stated that:

> There is no prospect of our finding a theory so penetrating that it will bring into rational order all or a large fraction of the heterogeneous elements of the landscape. There is no prospect of our finding such a theory, that is to say, unless it is of a mystical kind, and so outside the pale of science.
>
> There must be . . . selection among regions to be described as well as selection of items of information to be included in regional or topographic descriptions. But the regionalist position provides no logically given criteria for selection, save the most general one that the region selected exist on the earth.

Hartshorne addressed the issue of criteria of significance and selection in terms of a reading of the classics, most specifically in the arguments of Alfred Hettner. The logical arguments necessary to respond to this concern were similar to those of the Southwest German school of neo-Kantian philosophy, associated with Wilhelm Windelband and Heinrich Rickert. One of the concerns of the neo-Kantians was to establish the conditions which make possible the knowledge of the historical individual (Rickert 1986; Willey 1978; Oakes 1987; Smith, this volume). Geographers used these arguments as support for their concern with the geographic individual, the specific place or region.

Neo-Kantian arguments of interest to geographers are those concerning idiographic and nomothetic concept formation. Hartshorne (1939), through Hettner, used arguments similar to those of Windelband (1980) and Rickert (1986) on idiographic and nomothetic concept formation to support the geographers' concern with the individual case. Hartshorne agreed with Hettner's criticism of the neo-Kantians' early attempts to use this distinction to divide the natural and the human sciences, attempts they later abandoned. For Windelband and Rickert as well as for the sociologist Max Weber, to construct concepts is to create knowledge. Knowledge derives from human interests.

The neo-Kantians posited two distinct ways of knowing associated with two different cognitive or theoretic interests: an interest in knowing the general laws which govern phenomena and an interest in the individuality of phenomena. The facts of experience are not given, but rather are constructed in relation to these cognitive interests. Reality may be understood as being like an irrational continuum made rational through the application of concepts. This claim is not an ontological one, e.g., that reality is irrational, but rather is a phenomenological claim that describes our experience of the real. Concepts give order to this irrational flow of experience. The object of idiographic concept formation is to achieve a complete-as-possible understanding of the individual case. Nomothetic concept formation has as its goal the understanding of what is common among phenomena (Burger 1976; Oakes 1987; Rickert 1986; Weber 1949).

The nomothetic and idiographic modes of concept formation both derive from cognitive interests (Zaret 1980). Both are modes of abstraction through which finite minds seek to create rational order out of an infinite reality. The concept that is created given these conditions must necessarily be selective, and the validity of the representation is thus dependent on the validity of the criteria of selection. Lawful or law-like generalizations provide the grounds for nomothetic concept formation. In the words of Karl Popper (1950), theories act as the "searchlights" that provide the means of "illuminating" certain types of objects and events as being significant. Such generalizations are inappropriate, however, as guides for the study of the individuality of phenomena. According to the neo-Kantians, we seek to know about individual phenomena because we value their individuality.

Idiographic concept formation is thus rooted in values. Questions of selection and significance are addressed through the principle of "value relevance" (Burger 1976; Zaret 1980). Values serve as the criteria for the construction of the individual object, or, in the realm of the geographic, for the construction of specific place or region. References to criteria of significance stated in terms such as the "significance to man" and "geographical" significance may be interpreted in relation to this principle of value relevance (Hartshorne 1959).

This brief outline of the nature of neo-Kantian concept formation allows us to clarify certain aspects of the geographic debate about the idiographic and the nomothetic. A commonly expressed view is that the idiographic refers to the study of unique objects. I have suggested that the idiographic refers instead to the theoretic or the cognitive interest in the particular. The neo-Kantians were not concerned with unique objects, but rather with why certain aspects of our experience are valued for their individuality. To discuss idiographic concept formation in terms of the uniqueness of objects in the world is to interpret neo-Kantianism as a form of naive realism.

There is a certain ambivalence on this issue in Hartshorne in that at times he refers to the fact that no two locations are exactly alike, which suggests a naive realism, but most of his discussion concerns the idea of region and place as mental constructs, which derive from a particular way of viewing the world. We can, for example, study the same area of the earth's surface either in terms of its individuality, or in terms of the characteristics it shares with other areas. The specific region refers to areas viewed in terms of their individuality, and the generic region refers to a type or category of area.

A second common misunderstanding considers the idiographic as purely factual description that does not involve generalization or causation. The distinction between the idiographic and the nomothetic implies little about the use of singular factual statements versus generalizations. Rather, as I have stated, it concerns the cognitive interest or goal of the analysis. If the goal is to understand an area of the world in terms of its individuality, then it is a form of idiographic concept formation. For example, Max Weber's ideal types represented a form of generalization that had as its goal the characterization of the individuality of cultural phenomena. It should not be understood as a generalization based upon a method of averaging, but rather as an analytic construct that characterizes a historically and culturally significant phenomenon (Burger 1976).

I can turn to Weber's work once more to suggest that idiographic concept formation involves causal judgments. He stressed the importance of singular causal relations, that is, the serial causality of events. His model of causation in the cultural sciences was based on arguments about cause in legal theory and in theories of statistics and probability (Turner and Factor 1981; Zaret 1980). Reference to such causal relations are also found in *The Nature*. In addressing the question of the criteria of significance, Hartshorne drew on Hettner to suggest that there are two criteria of geographical significance for choosing what to include in a regional study: one is the spatial contiguity of phenomena and the other is the causal connectedness of these phenomena. Hartshorne (1939, 240) translated Hettner as stating that:

> The second condition is the causal connection between the different realms of nature and their different phenomena united at one place. Phenomena which lack such a connection with the other phenomena of the same place, or whose connection we do not recognize, do not belong in geographic study. Qualified and needed for such a study are the facts of the earth's surfaces which are locally different and whose local differences are significant for other kinds of phenomena, or, as it has been put, are geographically efficacious.

Hartshorne and Hettner saw geography as both idiographic and nomothetic. Within neo-Kantianism, these modes of concept formation are seen as equally valid. Both modes involve generalizations and causal judgments.

Rickert held that the values that guide idiographic concept formation possess the same universality as laws of nature. For Weber, values lack such universality

(Zaret 1980). Thus idiographic concept formation for Weber was relative to a point of view. This degree of subjectivity associated with the idiographic study of specific place and region was acknowledged by Hartshorne. It was also the quality of regional study criticized by Leighly (1937), and later condemned by the spatial analysts as being inappropriate for a science of geography (Schaefer 1953; Harvey 1969). The center of the dispute for the spatial analysts was not Hartshorne's interpretation of the classics, but rather the epistemological grounds upon which some of the classical texts were based.

The spatial analysts drew apart the idiographic and nomothetic and redefined them in terms of the unscientific study of the unique and the scientific search for lawful generalization. *The Nature* became the symbol of the study of the unique, and it still bears this emblem.

## Going beyond *The Nature* in the Study of Specific Place

I have suggested reasons why *The Nature* has not played a more prominent role in recent discussions of place and region in modern geography. I have argued that the grounds for methodological debate have shifted since the time of Hartshorne's arguments and that these shifts have not been favorable to the type of argument that he produced. Also, the prevailing image of *The Nature* was a relatively dimensionless one, based on an oversimplification of Hartshorne's argument.

The modern value of this classical work in American geography is found in its commitment to the importance of place and region for a science of geography (Agnew; Sack, this volume). Hartshorne's recognition, through Hettner, of the inability of a decentered, nomothetic perspective to fully capture this specificity led him to idiographic concept formation as a model for the objective study of the particular or the individual case. The role of value judgments in this form of concept formation made the study of specific place and region more vulnerable to the criticisms raised about the objective character of a chorological science. One means of minimizing the impact of such criticism is to emphasize the study of the relatively objective facts of place and region. This emphasis was not a consequence of neo-Kantianism, as is illustrated by the sophisticated understanding of meaning, language and action that has been a part of that tradition (most notably in the work of Max Weber and Ernst Cassirer). From our present vantage point, the combination of neo-Kantian and positivistic arguments in *The Nature* seems limiting in the attempt to understand the role of place and region in modern life.

Geographers in the late-twentieth century can easily understand Hartshorne's concern for identifying what is distinctive about a geographic view and presenting it in terms of the logic of science. But we wish to go beyond this concern. We recognize that, in order to understand the role of place and region in social life, we need to explore not only the relatively objective reality of place, but also its subjective reality. We seek to understand the many layers of the meaning of place from both the insider's and the outsider's view. We construct narratives of place and region that draw from both the objective reality of public facts about places as well as the subjective reality of the actor's sense of being in place. We are concerned with areal variation in the modern world, as well as with the moral geography of modernity. These concerns go beyond those of Hartshorne in terms of the subject matter associated with geography. They are consistent, however, with the search for appropriate criteria of selection and significance that motivates many of the arguments found in *The Nature* and with the type of close, logical scrutiny that is the hallmark of this work (Stoddart, this volume).

The current contextual model in the history of geographic thought does not require us to abandon the sense of the importance of building on the past. We may no longer seek to invoke the authority of the authors of classical geographical texts, but is to our advantage to continually seek a better understanding of our projects in relation to theirs.

## *The Nature* in Perspective

In the following essays, the authors begin the process of re-engaging *The Nature*, both as a symbol of mid-century American geography and as a rich repository of arguments concerning the study of place and region. *The Nature* has had this dual existence seemingly from the time of its publication. Hartshorne described his work as a critical analysis of current methodological themes in terms of a past methodological literature. It often has been used by protagonists and antagonists alike, however, as a dogmatic manifesto. Both aspects of the text's "life history" are addressed here.

The themes of the essays intersect at numerous points. For example, most of the authors consider the inward-looking quality of *The Nature*. Also, several authors recognize the often-neglected American roots of the methodological orientation of *The Nature*, as well as Hartshorne's mixture of neo-Kantian and positivistic epistemological arguments. The common appreciation of the enormous influence of *The Nature* is evident in the tendency of the authors to use the text as a basis for explaining particular directions in American geography as well as for speculation concerning paths not taken. The order of the manuscripts reflects in part this overlapping of themes and in part a rough chronological ordering that moves from historical context, to reception, to modern relevance. I shall identify below some of the specific themes raised by the authors.

The first two essays consider the German and the American methodological debates of the early twentieth century. Commentators on *The Nature* traditionally have taken for granted its Germanic influences, most notably that of Alfred Hettner. Indeed, the entire American chorological tradition has been described as the intellectual descendant of German as opposed to French or English regional geography. Thomas Elkins helps set the context for understanding this relationship by providing a view of methodological debate in German geography between the Wars, and Karl Butzer questions the prevailing conceptions of the influence of German geographers on their American counterparts.

Elkins presents an overview of some of the main intellectual currents in German methodological debate, as well as a description of the institutional context that helped shape the debate. He demonstrates the intellectual vigor with which German geographers pursued methodological topics through analyses of debates involving Alfred Hettner, Otto Schlüter and Hans Spethman. It has become part of the folklore of American geography that the intensity of methodological interest in the early twentieth century was a result of the insecure position of geography in American universities, but Elkins hypothesizes that the exact opposite was the case in Germany. That is, methodological debate was intense precisely because of geography's well-established position within the university. Its intensity and occasional excess were facilitated by the splendid isolation of the German professor, or, to use the terms of the intellectual historian Fritz Ringer (1969), the isolation and prestige of Germany's intellectual "mandarins." Elkins outlines the degree to which National Socialism changed this set of institutional arrangements.

Butzer challenges the neat, alliterative, intellectual pairing of the German and the American geographers: Sauer and Schlüter, Hartshorne and Hettner. He suggests that Hartshorne's use of the methodological writings of German geographers, most specifically those of Alfred Hettner, led to an overemphasis of the more didactic, integrative concerns of regional geography as opposed to the more scientific concern with the causal interconnections between human cultures and the natural environment. He argues that Sauer more clearly captured the full spirit of German geography of the late nineteenth and early twentieth century with his greater emphasis on the role of the environment. This conclusion is part of a more general argument about the American roots of many of the themes found in *The Nature*. For Butzer, an important source of Hartshorne's methodological orientation is found in the rift in American geography between the Berkeley and Midwestern schools of thought (Porter 1978).

Fred Lukermann examines methodological debate in American geography during the period between the publication of *The Nature* and the *Perspective on the Nature of Geography* (1959). He describes Hartshorne's view of geography as derivative from Hettner's "liberal positivism," as opposed to neo-Kantianism. He thus characterizes the attacks on Hartshorne by Schaefer and the spatial analysts as being "familial." The connection between neo-Kantianism and positivism is complex, both logically and historically (Giedymin 1975; Willey 1978). Resolution of the neo-Kantian versus positivistic basis of Hartshorne's arguments depends, in part, on which themes of *The Nature* are at issue. Lukermann's concern with a hermeneutic tradition in historical/cultural geography highlights the positivistic tendencies of *The Nature*. He suggests that the silence of the cultural/historical geographers concerning *The Nature* is indicative of a greater intellectual divide in American geography than that associated with the spatial analysts' reaction to Hartshorne.

The tendency by geographers to cast Hartshorne as either hero or villain in modern American geography is clearly represented in the narratives of Geoffrey Martin and Neil Smith. Martin's essay concerns the Hartshorne-Schaefer controversy, a dispute of almost mythical proportions in the modern history of American geography. Through his analysis of unpublished and published documents, Martin provides us with a wealth of historical information concerning the relation between the two antagonists and the events surrounding the publication of their methodological statements. His reportage has an additional, illocutionary impact that derives from the introduction of historical specificities into the discussion of mythologized events. Martin's description and evaluation of the events surrounding this episode in American geography, especially his portrayal of spatial analysts in the uncharacteristic role of loyal sentimentalists, will probably add fuel to the dying embers of this controversy (Billinge et al. 1984).

This inflammatory prospect is given some immediate support in Neil Smith's essay. Smith does not revive Schaefer's arguments, and is not about to argue for geography as a spatial science. His criticism of Hartshorne bears a resemblance to that of the spatial analysts, however, in his concern with Hartshorne's "exceptionalism" and its hypothesized consequence, the isolation of geography from the social sciences. He develops the metaphor of a "museum" to describe the isolation of geography from "mainstream" social thought. In part, this isolation is a function of what Smith sees as Hartshorne's absolute conception of space, a criticism found in Harvey's (1969) positivistic critique of chorology. Whereas Harvey's argument derives from concerns over uniqueness, Smith's develops more generally from the sense of a distinctly spatial perspective that he associates with Kant. Thus, like Schaefer, Smith describes a Kantian legacy in geography, associates its Anglo-American variant with Hartshorne, and blames both in different degrees for the intellectual isolation of the field. Unlike Schaefer, however, Smith gives a multiple dimensionality to Kantianism and neo-Kantianism, and directs geographers toward what he describes as critical social theory rather than to spatial science.

The concluding two essays consider the re-emergence of place and region as concepts central to modern, theoretical geography and seek to understand the relation of Hartshorne's work to this interest. John Agnew comments on the intersection of structuration theory and transcendental realism that has generated much of the recent discussion of the study of place and of areal variation. *The Nature* has been virtually ignored by those seeking to connect the study of areal variation with social theory. Agnew seeks to redress this apparent lack, while at the same time noting the differences between this recent theoretical work and Hartshorne's chorological perspective. He illustrates a realist, theoretical understanding of place through a study of Italian electoral politics, and thus gives the reader a chance to judge the degree of overlap between the two approaches.

In the concluding essay, Robert Sack outlines the contours of modern theoretical debate and captures many of the complexities involved in the attempt to situate Hartshorne's chorology within it. He resists the common tendency to caricature Hartshorne's arguments, and in doing so identifies some of the theoretical, epistemological and ontological ambiguities associated with a modern reading of *The Nature*. He describes chorology as integrative in its attempt to draw together in the study of place many of the important relations identified in modern social theories, but as nonetheless restrictive in terms of the type of phenomena studied. Hartshorne sought an epistemological distancing that we associate with a theoretical attitude, but not one that loses sight of the individuality of places. Ontologically, Hartshorne leaves us with the puzzle of the conceptual versus the material quality of place and region. In Sack's interpretation, Hartshorne's arguments become an important, yet historically specific and limited, contribution to the now expanding body of work by geographers on the role of place in the study of human society and culture.

The variety of responses to *The Nature* offered in these pages illuminate many aspects of the text and its context. They also highlight the fractured and uneven intellectual terrain of modern geography. In this sense, the essays in this volume should also be read as "a critical survey of current thought in light of the past."

## Note

1. A shorter version of this essay was presented at the President's Plenary Session, Meeting of the Association of American Geographers, Baltimore, Maryland, March 1989.

## References

Alexander, Jeffrey C. 1989. Sociology and discourse: On the centrality of the classics. In *Structure and meaning: Relinking classical sociology*, ed. J. C. Alexander, pp. 8–67. New York: Columbia University Press.

Berdoulay, Vincent. 1981. The contextual approach. In *Geography, ideology and social concern*, ed. D. R. Stoddart, pp. 8–16. Oxford: Basil Blackwell.

Billinge, Mark; Gregory, Derek; and Martin, Ron. 1984. *Recollections of a revolution: Geography as spatial science*. London: Macmillan.

Burger, Thomas. 1976. *Max Weber's theory of concept formation: History, laws, and ideal types*. Durham, NC: Duke University Press.

Cosgrove, Denis. 1985. *Social formation and symbolic landscape*. Totowa, NJ: Barnes and Noble.

Dear, Michael. 1988. The postmodern challenge: Reconstructing human geography. *Transactions of the Institute of British Geographers* 13:262–74.

Entrikin, J. Nicholas. 1981. Philosophical issues in the scientific study of regions. In *Geography and the urban environment: Vol. 4*, ed. D. T. Herbert and R. J. Johnston, pp. 1–27. Chichester, England: John Wiley.

Giedymin, J. 1975. Antipositivism in contemporary philosophy of social science. *British Journal for the Philosophy of Science* 26:275–301.

Gregory, Derek. 1978. *Ideology, science and human geography*. London: Hutchinson.

Guelke, Leonard. 1978. Geography and logical positivism. In *Geography and the urban environment: Vol. 1*, ed. D. T. Herbert and R. J. Johnston, pp. 35–61. Chichester, England: John Wiley.

**Hartshorne, Richard.** 1939. The nature of geography: A critical survey of current thought in light of the past. *Annals of the Association of American Geographers* 29:171–658. (Reprinted as *The Nature of Geography*, Lancaster, PA: Association of American Geographers, 1939.) All page references in this paper are to this reprinted edition.

———. 1955. "Exceptionalism in Geography" re-examined. *Annals of the Association of American Geographers* 45:205–44.

———. 1959. *Perspective on the nature of geography*. Chicago: Rand McNally.

**Harvey, David.** 1969. *Explanation in geography*. New York: St. Martin's Press.

**Johnston, R. J.** 1979. *Geography and geographers: Anglo-American human geography since 1945*. New York: John Wiley.

**Leighly, John.** 1937. Some comments on contemporary geographic method. *Annals of the Association of American Geographers* 27:125–41.

**Myres, J. L.** 1940. Review of *The Nature of Geography. Geographical Journal* 95:398–99.

**Oakes, Guy.** 1987. Weber and the Southwest German School: The genesis of the concept of the historical individual. In *Max Weber and His Contemporaries*, ed. W. Mommsen and J. Osterhammel, pp. 434–46. London: Allen Unwin.

**Popper, Karl.** 1950. *The open society and its enemies*. Princeton, NJ: Princeton University Press.

**Porter, P. W.** 1978. Geography as human ecology: A decade of progress in a quarter century. *American Behavioral Scientist* 22:15–39.

**Richardson, Miles.** 1984. Place and culture: A final note. In *Place: Experience and symbol*, ed. M. Richardson, pp. 63–67. Baton Rouge: Geoscience Publications of the Department of Geography and Anthropology, Louisiana State University.

**Rickert, Heinrich.** 1986. *The limits of concept formation in natural science: A logical introduction to the historical sciences*, trans. Guy Oakes. Cambridge: Cambridge University Press.

**Ringer, Fritz.** 1969. *The decline of the German mandarins: The German academic community, 1890–1933*. Cambridge, MA: Harvard University Press.

**Sauer, Carl O.** 1941. Foreword to Historical Geography. *Annals of the Association of American Geographers* 31:1–24.

**Schaefer, F. K.** 1953. Exceptionalism in geography: A methodological examination. *Annals of the Association of American Geographers* 43:226–49.

**Soja, Edward.** 1989. *Postmodern geographies: The reassertion of space in critical social theory*. London: Verso.

**Stoddart, David.** 1986. *On geography and its history*. Oxford: Basil Blackwell.

**Tuan, Yi-Fu.** 1974. Space and place: Humanistic perspective. *Progress in Human Geography* 6:213–52.

**Turner, Stephen, and Factor, Regis.** 1981. Objective possibility and adequate causation in Max Weber's methodological writings. *Sociological Review* 29:5–28.

**Weber, Max.** 1949. *The methodology of the social sciences*, trans. and ed. Edward Shils and Henry Finch. New York: Free Press.

**Willey, Thomas.** 1978. *Back to Kant: The revival of Kantianism in German social and historical thought, 1860–1914*. Detroit: Wayne State University.

**Windelband, Wilhelm.** 1980. History and natural science. *History and Theory* 19:165–85.

**Zaret, David.** 1980. From Weber to Parsons and Schutz: The eclipse of history in modern social theory. *American Journal of Sociology* 85:1180–1201.

# Human and Regional Geography in the German-Speaking Lands in the First Forty Years of the Twentieth Century

T. H. ELKINS

Honorary Professor of Geography, University of Sussex,
4 Arthur Garrard Close, St. Bernard's Road, Oxford OX2 6EU, England

Methodological discussion by German-speaking geographers in the first forty years of the twentieth century was of a volume and vigor unparalleled in Britain or North America. One reason for this was that academic geography was solidly established in the universities, with roots reaching back into the nineteenth century. In an unpublished paper contributed in 1988 to the IGU/IUHPS Commission/Working Group on the History of Geographical Thought, G. Sandner (1988) established that, apart from Ritter's 1825 Berlin chair (which, however, lay vacant for fifteen years after his death), the great period for the establishment of chairs was 1870–80, with eleven creations; by 1914, there were chairs at twenty-three universities in what is now the German Federal Republic and the German Democratic Republic. By 1933, this same area had thirty full chairs (plus one vacant) in twenty-eight institutes. There were four more in four institutes situated in German territory lost in 1945, and another ten in nine institutes in German-speaking universities that, then as now, lay outside the boundaries of Germany (Brogatio 1988). For Britain, by contrast, Chisholm (1908) writes of only a single chair at University College, London, although there were non-professorial appointments at a handful of other universities. The first full Honours School of Geography was established in Liverpool as late as 1917, although others were to follow in the interwar years. For the U.S. in 1906, Chisholm gives eleven professors or other teachers of geography distributed over seven universities or other institutions of higher education. In other words, the German-speaking lands were far in advance of Britain and the U.S. in the number of geographical chairs and the length of time for which they had been established; there were many potential participants in the debate, with, behind them, a respectable length of time for ideas to mature.

It is therefore understandable that Hartshorne's residence in 1938–39 in the German language area, principally in Vienna, coupled with his own command of the German language and his obvious interest in the German methodological discussion, meant that German-language literature figured prominently in *The Nature of Geography*. It cannot too strongly be stressed, however, that Hartshorne did not set out to write a history or a critique of German geography; he was concerned with the development of geographical thought as a whole, independent of national considerations. He took from the voluminous German methodological discussion what he felt to be relevant to his purpose, rejecting what he judged to be ephemeral or of purely local significance. For example, he declined to discuss one set of trends in the German geography of his day, as represented by the "new geography" of Spethmann, Muris and others, specifically stating: "Because the new *Weltanschauung* involved is limited to German geographers, the writer has not considered it necessary to examine it in this study" (1939, 138). There does, however, appear to be some utility and interest in outlining the intellectual background against which the book was researched. In so doing, there will be no attempt to criticize the German methodological literature, or Hartshorne's use of it; all that is modestly attempted here is to provide, for the benefit of those unfamiliar with the German literature, a necessarily brief outline of the intellectual environment in the midst of which Hartshorne undertook his research. If there are conclusions to be drawn, lessons to be learned, they will remain implicit rather than explicit.

# The Social Structure of German Geography

Geography since World War II, as elsewhere, has not lacked its mandarins, its Trolls and Gerasimovs, but their penchant has tended to be managerial or political, bestowing professional chairs or research grants upon those of whom they approve. What by contrast is likely to startle present day geographers about Germany in the first forty years of the century is not intrigue over posts so much as vehemence, not to say violence, with which academic controversy was carried on in the German language-area in the first forty years of the twentieth century; it is interesting to consider why this should have been so. The lonely pre-eminence of the typical professor may have been one factor. When, after a lengthy period of subservient apprenticeship, a geographer acquired the position of full professor (*Ordinarius*) he thereafter reigned supreme, in total and undisputed control of his university institute. Moreover, these institutes were extremely small; it was not impossible for an institute to consist only of the professor and a single assistant, totally dependent on him for advancement to the coveted goal of placement in a university chair. Appointment to supplementary chairs on a personal basis was rare, while the occasional private docent was also dependent on the *Ordinarius* for advancement and for the teaching opportunities from which he drew his slender income. The compilations of Sandner (1988) for 1914 and figures derived from Brogatio (1988) for 1933 give only Berlin as having more than one full chair, and this higher level of staffing could be precariously maintained only because of the existence of chairs at specialist institutes for historical geography, colonial geography and economic geography (in the Institute and Museum for Maritime Studies), in addition to the *Ordinarius* in the Geographical Institute (Albrecht Penck 1906–26, Norbert Krebs 1927–43, with a period of renewed service in 1946–47). Elsewhere only Vienna had two full chairs.

In principle, only those in possession of the *Habilitation*, a kind of higher doctorate opening the way to a university chair, could teach or examine. At this period most of them were full professors, and many of them retained their position of unchallenged eminence in the same institute for long periods; Hettner (1859–1941) was at Heidelberg from 1899 to 1928, Passarge at Hamburg, first at the Colonial Institute, subsequently at the University, from 1908 to 1935. Perhaps the most extraordinary achievement was that of Schlüter, who held the chair at Halle from 1911 until formal retirement in 1938 but, owing to the exigencies of the war and postwar period, was intermittently in charge until 1951, a period of forty years. These professors had ample time to become accustomed to command.

Another element of explanation may be looked for in the changing relationship of the academic to society. Almost without exception, professors of geography came from families that were representative neither of the industrial bourgeoisie (a relatively late developer in Germany) nor of the working class (secondary and higher education had to be paid for) but overwhelmingly from professional families, often with a record of serving the state over some generations. So among the geographers mentioned in this chapter, Hettner was the son of an art historian and Director of the Saxon Royal Museum of Antiquities, Schlüter's father was a lawyer, as were many of his mother's family, Passarge's father was a judge, Lautensach's a scholarly *Gymnasium* teacher, Rühl's an ancient historian, while his grandfather on the mother's side was the famous anatomist Henle. Interestingly, the Jewish Alfred Rühl had, in spite of his academically impeccable antecedents, arguably the most questioning and skeptical mind of his geographical generation (one must not forget Wittfogel, 1929), but otherwise possession of a chair was not likely to change inherited attitudes to society, but rather to increase self-esteem. Full professors were civil servants (*Beamte*) who, until the Hitler era, could expect security of tenure and a substantial pension on retirement. Many were honored by the state; among others, Hettner, Philippson, and Albrecht Penck bore the title *Geheimrat* (Privy Councillor), and were formally so addressed.

Intellectually unchallenged within their own little kingdoms, it is not surprising that some became dogmatic in the extreme. There were undoubtedly other influences contributing to the bitterness of debate. A feeling of personal insecurity has been suggested; in Prussia especially, the intellectuals had been almost a distinct class of (for the most part) servants of the state; these mandarins were now feeling threatened by the rise of both the industrial and commercial bourgeoisie and the proletariat in the exploding cities that were transforming the land (Ringer 1969). War, inflation and world economic depression removed old certainties, while the sharpening political battle could not be excluded from the universities, especially when joined with an anti-semitism that did not begin with the Hitler era.

Certain of the professorial attacks on colleagues bordered on paranoia, or even crossed into that category. The record was undoubtedly held by Siegfried Passarge (1867–1958), who had various victims. His attack on Hermann Lautensach (1886–1971), at the end of the 1920s, is generally held to have been an indirect attack on Lautensach's teacher, Albrecht Penck (Beck 1987). In particular, Passarge mounted a sequence of assaults questioning the academic integrity of Carl Troll (1899–1975) that began, apparently, with a trivial dispute over a review by Troll of one of Passarge's books in 1928, and continued even after the Second World War, when Troll was the undisputed leader of German geography. Passarge's attacks were even more dangerous in that, for a critical period in 1933–34, he was *Reichsobmann für Geographie* (Nazi chief for geography), but, fortunately, in 1934 Passarge's extremism led to his removal from office by the Nazis for applying his customary bullying tactics to de Martonne as head of the International Geographical Union at a moment when it was not policy to be beastly to the French (letter from Passarge to de Martonne, 18 October 1933, communicated by Sandner 1989). His departure did not entirely help Troll, as Passarge was succeeded by Mortensen (1894–1964), another Nazi professor who had collaborated with Passarge in his attacks, but who at least exhibited greater prudence. Even contributions to the methodological debate by such respected figures as Alfred Hettner, not to speak of his opponents in the "new geography" of the 1920s and 30s that will be discussed below, were sometimes of a degree of acerbity that would have been unusual in the Britain and America of the day.

This account will mainly center upon two figures. Alfred Hettner was preeminent as a proponent and defender of what he saw as the main stream of German geographical thought. The second figure was Otto Schlüter (1872–1959), who some contemporaries regarded as the more significant thinker, a view that has tended to gain in substance in more recent times. Limitations of space will prevent examination of alternative approaches to the landscape view of geography put forward by Passarge and Ewald Banse (1883–1953). Another regretted omission is the treatment of the ideas of "holism," "harmony," and "rhythm" in geography as favored by Gradmann (1865–1950), Granö (1882–1956), Volz (1870–1958 (1926, 1932)) and others. Although not dominant schools of thought, they were sufficiently distinctive to be attacked by Hettner and to be of some retraction to the more mystical National Socialist geographers. The same constraints prevent examination of a number of specialist fields within human geography, but it should be noted that one of these, political geography/geopolitics, has received considerable English-language coverage in recent years (Parker 1985, 1988; Heske 1986, 1987; Bassin 1987; Paterson 1987). There is also an English-language account of the pioneering work in economic (and indeed psychological) geography of Alfred Rühl (1882–1935), who, in a time of Nazi racist oppression, died in unexplained circumstances in Switzerland (Harke

1988). The various fields of physical geography will also not be examined, as lying beyond the competence of the author. It is unfortunate that space cannot be found for a detailed examination of Albrecht Penck (1858–1945) who, in addition to being a geomorphologist of international stature, contributed to the discussion on the nature of geography (1928). It should, however, be noted that it was customary for German geographers of the period to qualify initially in physical geography, a circumstance not without influence on the methodological approach even of those who went on to specialize in human aspects of the subject. To give one example, Rühl began his career as a geomorphologist and marine geographer; his mentor Penck was not pleased when he rejected environmental determinism as significant in economic geography (Harke 1988). From a later geographical generation, Ernst Plewe (1907–86) took his doctorate in the methodology of geography, but felt obliged to balance this by selecting a topic from physical geography for his *Habilitationsschrift* (Wardenga 1989). There being, normally, only one *Ordinarius* for each institute, it was essential at least for teaching, if not for research, to straddle the physical-human divide, so there were many contributions to methodological discussion from geographers primarily active in physical geography.

# Hettner and Regional Geography

At the end of the nineteenth century, German geographers (as geographers elsewhere, and neither for the first time nor the last), were concerned both with the position of their subject within the system of disciplines as a whole, and with its internal methodological problems. In spite of the historic contributions of von Humboldt, Ritter, Ratzel and von Richtofen, there was a feeling that geography had strayed, allowing its claims to spread over too many fields of knowledge, which were currently being reclaimed by a range of intellectually more coherent systematic disciplines. The very existence of the subject seemed to be in danger. It was felt that there was a need to identify a body of factual matter, or alternatively to identify an approach, that could be regarded as specifically geographical, immune from attacks from outside the discipline. At the same time, there was a feeling that geography should not concern itself primarily with the physical environment, but should maintain some degree of balance between both physical and human aspects of the subject. For some this meant a holistic approach in the manner reaching back to von Humboldt and Ritter (Schultz 1980).

Hettner's views on the position of geography in relation to the body of science as a whole were generally accepted by his contemporaries, and would in the main be accepted today. Hettner held that there were really no sciences, but only one science, one body of knowledge, and that divisions between disciplines were arbitrary. There was to be no "Great Wall of China" separating disciplines; crossing was permissible, but some agreed boundaries were operationally necessary, so that each discipline could have a distinct field within which it could operate according to its individual rules and objectives of study (1927). He rejected the notion that geography could be an all-embracing "science of the planet earth"; the latter was properly the object of study of a range of systematic disciplines, operating according to their own specific rules. While the objects treated by these disciplines sometimes interacted, any unified study of total global interactions, after the manner nobly attempted in von Humboldt's *Kosmos* (1845–62) was beyond intellectual possibility. Clearly geography had an interest in the spatial variations and distributions of individual phenomena, but this concern could not lie at the heart of the subject, because it was inevitably shared with the various systematic sciences. Geography also had to be rescued from being a complex of spatial aspects of other disciplines; it had to be given its own independent existence. Also rejected as unsatisfactory was the influential definition by von Richtofen (1833–1905) which held geography to be: "The science of the earth's surface and of the objects and phenomena in causal relationship with it" (1927, 122). This, Hettner felt, encouraged the "whole earth" approach which he rejected.

Hettner clearly thought that, in its retreat from what was felt to be the excessively humanity-centered and teleological approach of Ritter, geography had moved too far towards attempting to be a natural-scientific discipline concerned with the distribution of forms and phenomena, and that this had led to the acceptance of definitions that ignored the nature of the subject as it had developed in the previous hundred years (1927). Wherein, then, did he see the true path to geographical understanding? This, he felt, was to be sought in a return to the roots of the modern geographical tradition as Hettner saw it, to Ritter without his teleology, with himself as "the cleanser of the temple" (Schultz 1980, 80):

> Geography is not the general discipline of the [planet] earth. . . . Approaches to the surface of the earth as a whole, that is without reference to its spatial differentiation, do not fully attain the status of geography, which is rather the discipline of the earth's surface . . . according to its continents, countries, regions and localities. The term *Länderkunde* [regional geography] better expresses this content of the discipline than the term *Erdkunde* [geography of the earth's surface], which to Ritter would have been an unutterable expression, but which has led recent writers on methodology to produce false theoretical formulations on the nature of geography. However it is necessary to go beyond the description of individual countries and regions to a concern with comparative regional geography (1927, 122–23).

Hettner, as already noted, accepted that geography must include the systematic study of particular phenomena over the world as a whole, but:

> The most important characteristic of the geographical approach is that it is chorological in nature, and from this derives its unity. Chorology, however, is not a method to be ranked with other methods of description or explanation. A "method," if the meaning of the word is not unwarrantably to be extended, provides the path to an objective; chorology is the objective itself, the subject matter of geography. It involves viewing terrestrial reality from the point of view of spatial distribution, as opposed to the systematic sciences, which views reality in terms of its material differentiation, and the historical sciences, which view it in terms of sequence in

time. The geographic approach can never be anything other than chorological . . . (1927, 122–23).

Central to chorology is the concept of the total causal relationships of an assemblage of phenomena at a certain place on the earth's surface, which causes each place to be considered as a whole and stamps it with individuality. Ubiquitous phenomena, which do not vary spatially, are of no interest to geography; the same is true of phenomena that do vary spatially, but do so randomly, with no interrelationship with other varying phenomena (1927, 217). Of geographical interest are:

> such phenomena of the earth's surface as vary from place to place and whose spatial variation is significant for other groups of phenomena which are . . . geographically significant. The objective of chorological interpretation is the recognition of the character of countries and regions through the understanding of the assemblages and interrelationships of the various realms of reality and their various manifestations, and the comprehension of the surface of the earth as a whole in its natural division into continents, countries, regions and localities (1927, 130) [in part from *Perspective*, 13].

There is an evident similarity between these views and those of Hartshorne as summarized in the *Nature* (460–68) and in ch. 2 of the *Perspective*, but to Hettner they were dogmatically regarded as excluding all others. "He who has not absorbed this [particular approach to geography] into his flesh and blood has not comprehended the spirit of geography" (1905, 276), or again "The geographer who does not cultivate the regional approach is always in danger of totally losing contact with the solid basis of the subject. Who does not understand this is not a true geographer" (1919, 23) [quotations from Schultz (1980, 85)]. On the other hand, this regard for the regional as the core of the subject does not mean a total attachment to the ideographic; laws were to be sought where possible and appropriate, as Hartshorne has claimed (1988) and, indeed, as the field research activities of Hettner and his students demonstrated.

Hettner's methodological views were repeatedly put forward in the pages of the *Geographische Zeitschrift*, the journal that he founded in 1895 and which he edited until he felt obliged to withdraw in the adverse racist and political climate of 1935. It would go too far to say that the journal acted as the "monitoring authority in relation to unconforming opinions" (Schultz 1980, 122), and that no departure from Hettner's views was allowed in the journal; variations on the central theme were permissible, but total rejection was not, at least when the political tone began to sharpen with the emergence of the "new geography" of the late 1920s (see below).

At the time when Hettner wrote, there was considerable discussion on the question of environmental determinism. In this respect, as in some others, Hettner's statements were sometimes contradictory. Partly, although not perhaps entirely, this was because his principal methodological statement was a summation of positions developed over thirty years, which lamentably lacks an index (Hettner 1927). In general, Hettner's approach to geographical theory must be placed firmly in the environmental-deterministic category, giving predominance to land-humanity relationships: "passing over the decisions of human will, we relate the facts of human geography to the conditions imposed by the nature of the land" (1927, 267, similarly 210). Of the scope of human geography, he wrote: "Health, hygiene, nourishment, clothing, dwelling, recreation, education and spiritual cultural objects are subjects of geographical study, as they are dependent on natural conditions. . . . [Human geography] "extends to most manifestations of human life, but only to the extent that they are in close interrelationship with the nature of the land, and can be apprehended as manifestations of the nature of the land" (1927, 150). He repeatedly refers to "chains of geographical facts" or "chains of causation" (*Ursachenreihe*). But it is equally possible to find, and within the same publication, directly contrary expressions. Speaking of economic geography:

> The land is always the objective of geographical explanation. It is however an incorrect application . . . of geographical principles, made by many workers, to limit economic geography to the geographical [natural] determining of economic phenomena. This is to pull the rug from under their feet, for influences as such can never be the subject of scientific treatment" (1927, 149).

Most controversial of all was Hettner's adherence to the so-called *Länderkundliche Schema* (regional-geographical model), in which was treated sequentially geological structure, surface morphology, climate, drainage, plant geography, animal geography, settlement, economy and trade, and population (Spethmann 1928; Dickinson 1969). The approach was not restricted to Hettner, being seen at its highest intellectual level in works such as Philippson's *Das Mittelmeergebiet* as well as in many textbooks of regional geography of very varying quality; the present writer has to confess to having used it himself in younger days (Elkins and Yates 1963). It has to be admitted that the unimaginative application of such a stereotyped *Länderkundliche Schema* could readily produce a rather unilluminating result. This was admitted by Hettner following criticism, who nevertheless stated that the sequence could justifiably be applied in the same order in many countries and localities (1931, 111). It is also evident that the *Schema* is inherently environmental-deterministic, whether this was appreciated by former practitioners or not; the ordained sequence of features ensures that the "physical basis" is inevitably "basic."

Hettner's response to this criticism would undoubtedly have been that the *Schema* was a didactic device, not a research programme. It has, in fact, been observed that neither Hettner's original research work in Latin America, nor the work of his doctoral students, fell within the compass of the regional geography that, in theory, he regarded as forming the crown of geography (Dickinson 1969). All of them conducted research work on a selected systematic topic, using such research methods as were appropriate, derived if necessary from a neighboring discipline. Recently it has been suggested that the whole of Hettner's engagement with chorology or regional geography was not a search for a research method but a search for a method of conveying information about the differentiation of the earth, basically didactic in purpose, and concerned essentially with how to organize and present existing knowledge in a rational fashion. The approach was not designed to replace systematic research but to stand alongside it, indeed providing geography's culmination (Wardenga 1987). This is confirmed in one of Hettner's last papers, written as a defense against criticism (Hettner 1931). If, then, Hettner's insistence on the chorological or regional approach has this limited end, the passion with which he defended it becomes astonishing.

## Schlüter and the Landscape Concept

Divergent views on the nature of geography between Hettner and Schlüter became apparent as early as the end of the nineteenth century.

Schlüter was inevitably familiar with Richtofen's definition of geography given as: "The science of the earth's surface and of the objects and phenomena in causal relationship with it." As von Richtofen's views developed, he became convinced that only the physical part of geography had an objective material substance, whereas organic-human aspects existed only in a dependent relationship to the physical (Lautensach 1952). Schlüter was to depart from the approach of his master in his extraordinarily precocious working out, as a young man in his 20s and 30s, of precisely formulated views on the methodology of geography that stayed with him for the whole of a long life.

A pillar of his position was the rejection of the notion of geography as a science of relationships and particularly of human geography as a study of nature-man relationships, which he felt to be not only deterministic but leading to misconceptions, although this was a view still put forward, in one version or another, by some geographers fifty years later (Martin 1951; Spate 1952).

It is greatly to the credit of von Richtofen that he encouraged Schlüter to develop his own ideas, irrespective of their conflict with his own. Schlüter's doctoral thesis on settlement in the Unstrut valley, carried out in his Leipzig period under the supervision of Adolf Kirchhoff (1838–1907), had already led him to observe the very different settlement patterns emerging from occupance by German and Slav settlers of contrasting culture in areas of similar physical constitution. In a very early paper completed in Berlin, but relating to his Halle researches, he wrote:

> It is a misunderstanding to believe that human-geographical phenomena owe their place in geography to geographical [natural] causation, in other words, that the object of study of human geography is the dependence of humanity upon natural conditions or the influence of natural conditions upon humanity. . . . With the emphasis on geographical [natural] determinism a quite different idea is introduced into geography. The objective is no longer geography itself, but relationships. From the large number of influences operating in a single case, one group [natural phenomena] is selected, leaving aside the others as geographically without interest. There is no doubt that valuable findings can be revealed in this way, but . . . this approach carries the danger of an intellectual bias and prejudice, leading to an effort

to attribute as much as possible to geographical [natural] relationships (Schlüter 1899, 66–67).

The last sentence is critical; natural determinism is to be avoided because it leads to error. As an example of this, and indeed of the manifest absurdity of natural-deterministic explanation, Schlüter criticizes the then customary explanation of the rise of Greek civilization in terms of a benign climate and the existence of sheltered bays backed by lowlands within which the culture of the various city-states could develop. He was undoubtedly correct in querying this explanation, in view of the fact that no comparable development could be traced in the succeeding 2000 years, although the environment remained effectively unchanged (Schlüter 1906a, 588).

The question, according to Schlüter, is to explain the emergence of particular cultural phenomena in particular parts of the earth.

> In such a treatment man and his cultural development is the initial focus of attention; geographical [natural] environment appears not as the creative force, but as something unalterable, setting limits (Schlüter 1919, 8).

Only when the state of cultural development is known can statements about the significance of natural environment be made:

> the influence of geographical conditions makes itself apparent in the way that active and creative man purposely or otherwise makes use of the substance and forces of the earth, unconsciously makes use of the opportunities of his environment, or purposely transforms them to his purpose. Departing from human activity, we can fully assess the extent of geographical [natural] "influence," while beginning with nature, land and territory, we cannot make what is truly human really comprehensible, but only throw light in a one-sided and oblique fashion (Schlüter 1928, in Paffen 1973, 320).

Again:

> A simple inventory of geographical [natural] influences will not suffice, we must include all factors which have really been operative. These however include the factors that spring from man himself and his culture. They are not to be regarded as a second group standing alongside the natural factors, for in them alone lies the natural element, whereas the relationship to the earth's surface can only come into consideration as providing modifying conditions (Schlüter 1913–14).

As Schlüter himself said, there is nothing very penetrating about such views, but they were not usual at the time. Nor was his contribution limited to the destruction of environmental determinism; his positive contribution rested on the notion of the morphology of landscape, physical and cultural, as the essential object of geographical investigation. A word of explanation is necessary here. The word *Landschaft* was often used by Hettner and others in the sense for which we would use "region." Schlüter's usage is nearer to the Anglo-American one, referring to the visible, material content of the earth's surface. Schlüter would have maintained that, by about the end of the nineteenth century, the objects of geomorphological study were clearly recognized, and at least some commonly accepted methods of their study and explanations worked out. His proposal was that the objects of the cultural landscape should be investigated from just the same descriptive and explanatory point of view.

His approach was clearly articulated even before the beginning of the twentieth century:

> As with physical geography, so must human geography begin with the concrete phenomena and seek to understand them in all ways. The field of research is not what is "determined" by the character of the land, but that character itself. . . . What [human geography] aspires to is the recognition of the forms and disposition of terrestrial phenomena, in so far as they may be perceived in the sense of their spatial distribution, their visible appearance and their palpable expression. To accomplish this task, the geographer must have a completely free hand in the search for explanation and must give proper attention to all types of causation, whether derived from the nature of the land or from the spirit of man (1899, 67).

Similarly, as expressed by one of the most creative of geographers of the interwar years, a man that Nazism was later to lose from Germany:

> Schlüter was the first to raise the landscape-forming activity of man to a methodological principle. Originating from the logical application of the physiognomic concept of landscape, he gave the geography of man a corporate substance for research, the cultural landscape, which can be investigated according to the same method as physical geography. This can be examined from the points of view of its morphology, physiology, and developmental history, just as the visible phenomena of nature in the build of the landscape. Between physical geography and the geography of man there is no longer a gap. Both are in the closest contact in terms of objects and methods. Thus in my opinion, Schlüter's physiognomic approach is a great gain for geography, although thereby the field of its enquiry is greatly narrowed (Waibel 1933, 199; tr. Dickinson 1969, with amendments).

Schlüter and his followers held that the focusing on the landscape as an object of description and of causal-genetic explanation gave geography its individuality, because this object of study was shared by no other discipline. In the process, the scope of geography was necessarily narrowed, but this was accepted as a virtue, whereas the chorological-regional approach was held to require, at least potentially, the inclusion of all human activities, thus being impossibly wide in scope. There was the further advantage that, methodologically, the gap between human and physical geography disappeared, since all geography was concerned with the description and explanation of landscape. Schlüter's approach also excluded the consideration of all spatially varying phenomena that do not imprint themselves upon the landscape; the state, for example, was seen as having only a very general, background, impact on features of the cultural landscape, passing into insignificance by comparison with the detailed and immediate impact of, for example, surface morphology and vegetation.

Hettner intermittently attacked Schlüter's ideas, mostly in the *Geographische Zeitschrift*, for over a quarter-century following the appearance of Schlüter's paper on the objectives of human geography (Schlüter 1906b), to which Hettner devoted four pages of what can only be described as a bruising attack on a younger and less firmly established colleague (1907–08, 628–82). Hettner's criticisms prior to 1927 may be regarded as summarized in his *Geographie*, published in that year, but he returned to the subject later (1929, 276–81). Schlüter, by contrast, appears to have had little appetite for direct controversy.

As set out in the *Geographie* (1927, 128–29), Hettner is fundamentally disturbed by Schlüter's rejection of chorology. A landscape approach he can accept only in terms of aesthetic geography, which indeed he regards as insufficiently developed by Banse and Younghusband (Hettner 1927; the book has no bibliography). With reference to the landscape approach, he writes:

> But geography as a whole cannot be so one-sided; for example, the soil cannot be interpreted simply in terms of its colour instead of its physical and chemical composition; climate cannot be restricted to the colour of the heavens and the cloud cover, while treatment of realm of plants and animals cannot exclude floristic and faunal differences, merely because they stand out only to a limited extent in the landscape (1927, 128).

Hettner particularly condemned what he considered to be the reduction of the human element in geography to what can be perceived:

> With the exclusion of the mental element, geography loses areas that have long been cultivated with particular assiduity, such as political geography, ethnic geography and effectively the geography of transport and trade, for which the study of visible transport means is no substitute (1927, 129).

In a comment that became notorious, Hettner accused the proponents of the landscape approach of formally excluding these items, but, in fact, finding themselves forced to "smuggle them in by the back door" in the course of explanation.

Apart from his methodological contributions, Schlüter's research work lay essentially in the field of settlement geography, especially of central Germany. He devoted much of the second half of his long academic career to a painstaking effort to reconstruct the landscape of central Europe at the onset of German settlement expansion in early-medieval times, publication of which was mainly posthumous (*Siedlungsräume Mitteleuropas in frühgeschichtlicher Zeit* 1952, 1953, 1958). Studies of the evolution of the cultural landscape and of the forms of rural settlement, of the type pioneered by Schlüter, became a prominent feature of German geography. As a research direction in difficult political circumstances, it had the advantage of being uncontroversial, indeed as having a patriotic element in dealing with the work of Germans in transforming the landscape within and beyond the boundaries of post-1918 Germany.

Although it is clear that Hartshorne's 1939 position was much closer to that of Hettner than to that of Schlüter, he recognized the significance of the landscapist approach, both in German geography and in other countries, devoting to it an extended and scholarly treatment (1939, 149–74, 189–236, 346–50). It was an approach that, through the work of Carl Sauer (1925) and pupils, has been enormously influential in geography in the United States. Looking backwards from the present day, when perceptual geography and the interpretation of landscape are subject to considerable attention, the contribution of Otto Schlüter can only gain in stature.

# The *Dynamische Länderkunde* Affair

From 1927, Dr. Hans Spethmann occupied the insecure position of *Privatdozent* in the University of Köln. Clearly a man of outstanding intellect, he was perhaps injudicious, from the viewpoint of career advancement, to tackle head-on the mandarins of his profession, notably Hettner. Spethmann's ideas would not today be thought to be at all out of the way. He wished to move on from what he regarded as a very static, descriptive, regional geography to one which took account of what he saw as the changing forces that actually produced the spatial variation of the land. These, in addition to the familiar surface morphology, climate and vegetation, he regarded as including forces derived from the development of technology; financial forces, crises and the level of business activity; forces derived from individual personality; political forces; religious forces; and imponderable forces, such as catastrophes. He regarded each of these forces as having had a distinct, if overlapping, spatial expression.

Spethmann first put forward his ideas in a relatively cautious "mentioning no names" paper in *Zeitschrift für Geopolitik* (1927), although the views criticized could easily be tracked down to their originators by any informed geographer, and were so identified by Hettner, who responded with an unusually brief if incisive rejoinder (1928). Spethmann's book *Dynamische Länderkunde* (1928) was another matter. Not only did he criticize the *Länderkundliche Schema* for its inherent geographical determinism, but he was imprudent enough to pick over the works of his seniors for what he conceived to be their errors in this respect. For example, he attacked the explanation by Machatschek, Hassinger and Hettner and others of the rise of Berlin as capital of the German Empire in terms of physical geography, referring to its central location within the North German Lowland, the convenient crossing of the Spree and its waterway connections to the Elbe and Oder. He rightly pointed out that the rise of Berlin was in fact determined by the decision of the Hohenzollerns, for entirely non-geographical reasons, to establish their residence there.

Spethmann could with every justification complain that he had been three times attacked in the *Geographische Zeitschrift* (Gradmann 1928; Hettner 1929; Philippson 1930) without being accorded any right of reply. But if Hettner could rally his allies, so could Spethmann (Muris 1930). He was also a doughty fighter on his own account. In his *Das länderkundliche Schema in der deutschen Geographie; Kämpfe um Fortschritt und Freiheit* (1931), he was unsporting enough to print letters he had exchanged with Hettner's supporters. To some extent, the debate can be interpreted as a battle between generations; it would not be the last time that a "new geography" would be used as a weapon by a rising generation against the "old men." There was also a political element, of young right-radicals against aging liberals; Muris (1884–1964), in particular, was a prominent Nazi geographer who was to go on to hold chairs in various of the special teachers' training colleges set up under the National-Socialist regime, and who appears not to have been re-employed after 1945. In retrospect, Spethmann can be seen as having given at least as good as he received in the argument, perhaps as having won it, but his Cologne *Ordinarius* Franz Thorbeke (1875–1945) was a Hettner student, and he never attained a university chair. In a sad little booklet published in 1938, "My farewell to geographical teaching" (*Mein Abscheid von der geographischen Lehrtätigkeit*), he announced his acceptance of the fact that he was not going to advance in the geographical world (he was to become an author of major importance in relation to aspects of the Ruhr industrial region). It is a commentary on the lack of total control by the Nazis over the German universities that a group of geographers opposed to Nazi policies (including Jewish or part-Jewish persons) could exercise such influence so late in the 1930s.

Hartshorne (1939, 138), as already noted, declined to interest himself in the Spethmann affair, as being specifically German rather than of general interest; it may well be that he had already caught a whiff of political involvement in the affair.

## Geography under National Socialism

For an overview of the work of German-speaking geographers in the period 1933–45, readers may conveniently refer to the admittedly incomplete, and only partially translated, survey by Troll (1949). Unlike the Marxist ideology imposed upon geography in part of Germany from c. 1950, National Socialism had no coherent intellectual structure, but some characteristics can be isolated from the programmatic statements of geographers sympathetic to the Nazi cause. Hettner had held that: "Geography . . . can only be based on the principle of a value-free science and on the independence and internal coherence of its subject matter" (1919, 15). The Nazi supporters would have none of this: "Away with misguided 'pure' science, away with the refusal to admit to values, away with equally misguided attempts at 'objectivity,'" to be replaced by concern with the politics of the *Volk* (Muris 1934, 44). Or: "Science unrelated to the needs of the *Volk* is fit only for the museum" (Mortensen 1934, 534).

Geography under National Socialism was essentially German-centered: "National geography is for us the whole of geography, looking with German eyes and from the German standpoint upon Germany and the world" (Schrepfer 1934, 63). This was a great period for the drawing of "German-centered" maps showing the "creative" contribution of German-speaking people to the cultural landscape of Europe, maps which looked very different to the non-German populations at the receiving end (e.g., Meynen 1935a). It was a human geography that had "race" as its central concept. The notion of a pure Aryan or Nordic *Volk* was, of course, a myth, but clearly it brought about, at least in principle, a significant shift in how geographers of National-Socialist persuasion looked at the land. If racial characteristics, including mental attributes, were deemed to be inherited and not to be modified by environment, then there was no room for the environmental determinism that had previously been so prominent. This was one of the reasons for the quarrel between the Nazi sympathizers and Hettner, who was denounced for the "materialism" of his allegedly geo-deterministic approach to geography, which ignored the vital and creative force of the race or *Volk*. "It is not the case that geography (*Raum*) influences and forms the *Volk*, but that the *Volk* forms the *Raum* (Meynen 1935b). At the same time all geographical publications were increasingly subjected to censorship, either directly from the official propaganda machine or editorially on grounds of prudence; the process has been examined in detail in respect of the *Geographische Zeitschrift* under Hettner's successor, Schmitthenner (1887–1957) (Sandner 1983).

It should perhaps be mentioned at this stage that not all of the rising generation of geographers docilely fell into line with the new right-radical persuasion; at least one of them, Ernst Plewe (1907–86), was prepared to prejudice his career (or risk a worse fate) to criticize characteristic ideas of the geographers orientated towards National Socialism, such as the substitution of Nazi "values" for scientific objectivity, or the notion of a specifically German geography (Plewe 1935).

Then came the Second World War, with many geographers absorbed into war service of one kind or another. There was little time or inclination for methodological debate. Curiously, when German-speaking geography reappears in the Federal Republic, Austria and Switzerland after the war, methodological discussion still centers on the same two topics: regional geography and landscape. The "theoretical-quantitative revolution" came late to Germany, with the outstanding contribution of Dietrich Bartels (1968). But that is another story.

## Note

1. All translated passages are by the author, unless otherwise stated. It is not everywhere possible to combine accuracy of translation with avoidance of the contemporary use of what would now be considered, at least by some, as sexist language. Items in square brackets within translated passages are by the author.

## References

For an extensive bibliography covering the development of German-speaking geography 1800–1970, see Schultz (1980). For German contributions on the history of German geography written 1945–80, also see Sandner (1988).

**Bartels, D.** 1968. *Zur wissenschaftstheoretischen Grundlagen einer Geographie des Menschen* (On the foundations provided by disciplinary theory for the study of human geography). Wiesbaden: Steiner (Erdkundliches Wissen 19).

**Bassin, M.** 1987. Race contra space; the conflict between German *Geopolitik* and National Socialism. *Political Geography Quarterly* 6:115–34.

**Beck, H.** 1974. Hermann Lautensach—furender Geograph in zwei Epochen. Ein Weg zur Länderkunde (Hermann Lautensach—outstanding geographer in two epochs. An approach to regional geography). *Stuttgarter Geographische Studien* 87:1–42.

———. 1982. *Grosse Geographen; Pioneere-Aussenseiter-Gelehrte* (Great Geographers; pioneers-outsiders-scholars). Berlin: Riemer.

———. 1987. Hermann Lautensach. *Stuttgarter Geographische Studien* 100.

**Brogatio, H.** 1988. Personal communication.

**Chisholm, G. G.** 1908. The meaning and scope of geography. *Scottish Geographical Magazine* 24:561–75.

**Dickinson, R. E.** 1969. *The makers of modern geography.* London: Routledge & Kegan Paul.

**Elkins, T. H., and Yates, E. M.** 1963. The South German scarplands in the vicinity of Tübingen. *Geography* 48:372–92.

**Gradmann, R.** 1924. Das harmonische Landschaftsbild (The harmonious representation of landscape). *Zeitschrift der Gesellschaft für Erdkunde zu Berlin,* Special issue out of series: 129–47, 333–37.

———. 1926. Harmonie und Rythmus in der Landschaft (Harmony and rhythm in the landscape). *Petermanns Geographische Mitteilungen* 72:23.

———. 1928. Dynamische Länderkunde (Dynamic regional geography). *Geographische Zeitschrift* 34:551–53.

**Granö, J. G.** 1935. Geographische Ganzheiten (The geographic wholes). *Petermanns Geographische Mitteilungen* NN:295–302.

**Harke, H.** 1988. Alfred Rühl 1882–1935. *Geographers: Biobibliographical Studies* 12:139–47.

**Hartshorne, R.** 1939. The nature of geography; a critical survey of current thought in the light of the past. *Annals of the Association of American Geographers* 29(3 and 4).

———. 1959. *Perspective on the nature of geography.* Association of American Geographers. 1960, London: Murray.

**Heske, H.** 1986. German geographical research in the Nazi period; a content analysis of the major geography journals, 1925–1945. *Political Geography Quarterly* 5:267–81.

———. 1987. Karl Haushofer; his role in German geopolitics and in Nazi politics. *Political Geography Quarterly* 6:135–44.

**Hettner, A.** 1905. Das System der Wissenschaften (The system of the sciences). *Preussische Jahrbuch* 122:251–77.

———. 1907–08. Methodologische Streifzüge (Methodological disputes). *Geographische Zeitschrift* 13:627–32, 694–99; and 14:561–68.

———. 1919. *Die Einheit der Geographie in Wissenschaft und Unterricht* (The unity of geography as an academic discipline and in teaching). Berlin: Geographische Abende im Zentral-Institute für Erziehung und Unterricht 1.

———. 1927. *Die Geographie; ihre Geschichte, ihr Wesen und ihre Methoden* (Geography; its history, nature and methods). Breslau: Hirt.

———. 1928. Neue Wege in der Länderkunde (New directions in regional geography). *Zeitschrift für Geopolitik* 5:273–75.

———. 1929. Methodische Zeit und Streitfragen. Neue Folge (Contemporary methodological disputes, new series). *Geographisches Zeitschrift* 35:264–86, 332–45.

———. 1931. Die Geographie als Wissenschaft und als Lehrfach (Geography as a research and teaching subject). *Geographischer Anzeiger* 32:107–17.

**Humboldt, A. von.** 1845–62. *Kosmos, Entwurth einer physischen Weltbeschreibung.* 5 vols. Stuttgart and Tübingen.

**Lautensach, H.** 1952. Otto Schlüters Bedeutung für die methodische Entwicklung der Geographie; eine kritischer Querschnitt durch ein Halbjahrhundert erdkundlicher Problemstellung in Deutschland (The significance of Otto Schlüter for the methodological development of geography; a critical cross-section of the approach to geographical problems in Germany over half a century). *Petermanns Geographische Mitteilungen* 96:219–31.

**Martin, A. F.** 1951. The necessity for determinism; a metaphysical problem confronting geographers. *Transactions, Institute of British Geographers* 17:1–11.

**Meynen, E.** 1935a. *Deutschland und Deutsches Reich. Sprachgebrauch und Begriffswesenheit des Wortes Deutschland* (Deutschland [Germany] and Deutsches Reich [German Empire]; linguistic usage and terminological essence of the word Deutschland). Leipzig: Brockhaus.

———. 1935b. Völkische Geographie. *Geographische Zeitschrift* 41:435–41.

**Mortensen, H.** 1934. Inweifern kann die Hochschulgeographie den Bendürsnissen der Schulgeographie und der allgemeine Volks bildung gerecht werden? (To what extent can university geography do justice to the needs of school geography and of national education in general?). *Geographischer Anzeiger* 35:532–45.

**Muris, O.** 1930. Der Streit um die 'Dynamische Länderkunde'; eine Stellungnahme von seiten der Schulgeographie (The *'Dynamische Landerkünde'* debate; an opinion from the direction of school geography). *Geographischer Anzeiger* 31:285–91.

———. 1934. *Erdkunde und nationalpolitische Erziehung* (Geography and national political education). Breslau: Hirt.

**Paffen, K.-H., ed.** 1973. *Das Wesen der Landschaft* (The nature of landscape). (Collected reprints of varying dates with an Introduction by the Editor.) Darmstadt: Wissenschaftliche Buchgesellschaft.

**Parker, G.** 1985. *Western geopolitical thought in the twentieth century.* London: Croom Helm.

———. 1988. *The geopolitics of domination.* London: Routledge.

**Passarge, S.** 1919 etc. *Die Grundlagen der Landschaftskunde* (Foundations of landscape geography). 3 vols. Hamburg: Freidricksen.

**Paterson, J. H.** 1987. German geopolitics reassessed. *Political Geography Quarterly* 6:107–14.

**Penck, A.** 1928. Neuere Geographie. *Zeitschrift der Gesellschaft für Erdkunde zu Berlin; Sonderband zur Hundertjahrfeier der Gesellschaft,* pp. 31–56. Berlin.

**Philippson, A.** 1930. Methodologische Bemerkungen zu Spethmanns dynamischer Länderkunde (Methodological comments on Spethmann's *Dynamischer Länderkunde*). *Geographische Zeitschrift* 36:1–16.

**Plewe, E.** 1935. Randbemerkungen zur geographischen Methodik (Marginal comments on geographical methodology). *Geographische Zeitschrift* 41:226–37.

**Ringer, F. K.** 1969. *The decline of the German mandarins: The German academic community, 1890–1933.* Cambridge, MA: Harvard University Press.

**Sandner, G.** 1983. Die "Geographische Zeitschrift" 1933–1944; eine Dokumentation über Zensur, Selbstzensur und Anpassungsdruck bei wissenschaftlichen Zeitschriften im Dritten Reich (The "Geographische Zeitschrift" 1933–44; a documentation of censorship, political pressure and editorial adaptation of scientific journals during the Third Reich). *Geographische Zeitschrift* 71:65–87, 127–49.

———. 1988. Recent advances in the history of German geography 1918–45; a progress report for the Federal Republic of Germany. *Geographische Zeitschrift* 76:120–33.

———. Personal communication, 1988.

**Sauer, C.** 1925. *Morphology of landscape.* University of California, *Publications in Geography* 2:19–54. Reprinted in *Land and life,* ed. J. Leighly. Berkeley: University of California Press, 1963.

**Schlüter, O.** 1899. Bemerkungen zur Siedlungsgeographie (Comments on settlement geography). *Geographische Zeitschrift* 5:65–84.

———. 1906a. Die leitenden Gesichtspunkte der Anthropogeographie, insbesondere der Lehre Friedrich Ratzels (Leading aspects of human geography, with special reference to the teaching of Friedrich Ratzel). *Archiv für Sozialwissenschaft und Sozialpolitik* 22:581–630.

———. 1906b. *Die Ziele der Geographie des Menschen* (The objectives of human geography). Munich and Berlin: Oldenbourg.

———. 1913–14. Die Erdkunde in ihrem Verhaltnis zu den Natur- und Geisteswissenschaften (Geography in relation to the natural sciences and the humanities). *Die Geisteswissenschaften* 1:283–89, 310–15. Reprinted with amendments in *Geographischer Anzeiger* 21(1920):145–52, 213–18.

———. 1919. *Die Stellung der Geographie des Menschen in der erdkundlichen Wissenschaft* (The place of human geography in the discipline of Geography). Berlin: Mittler.

————. 1928. Die analytische Geographie der Kulturlandschaft erläutert am Beispiel der Brücken (The analytical geography of the cultural landscape, as illustrated by the example of bridges). *Zeitschrift der Gesellschaft für Erdkunde zu Berlin; Sonderband zur Hundertjahrfeier der Gesellschaft*, Berlin, pp. 388–411.

Schrepfer, H. 1934. Einheit und Aufgabe der Geographie als Wissenschaft (Unity and tasks of geography as a discipline). In *Die Geographie vor Neuen Aufgaben*, ed. J. Petersen and H. Schrepfer, pp. 61–86. Frankfurt am Main: Diesterweg.

Schultz, H.-D. 1980. *Die deutschsprachige Geographie von 1800 bis 1970; ein Beitrag zur Geschichte ihrer Methodologie* (German-speaking geography 1800-1970; a contribution to its methodological history). Abhandlungen des geographischen Institutus—Anthropogeographie 29. Berlin: Free University.

Spate, O. H. K. 1952. Toynbee and Huntington; a study in determinism. *Geographical Journal* 118:402–28.

Spethmann, H. 1927. Neue Wege in der Länderkunde (New directions in regional geography). *Zeitschrift für Geopolitik* 4:989–98.

————. 1928. *Dynamische Länderkunde* (Dynamic regional geography). Breslau: Hirt.

————. 1931. *Das länderkundliche Schema in der deutschen Geographie; Kämpfe um Fortschritt und Freiheit* (The regional-geographical model in German geography; struggles over progress and freedom). Berlin: Hobbing.

————. 1938. *Mein Abschied von der geographischen Lehrtätigkeit* (My farewell to geographical teaching). Berlin: Schmidt.

Troll, Carl. 1947. Die geographische Wissenschaft in Deutschland in den Jahren 1933 bis 1945; eine Kritik und Rechtfertigung. *Erdkunde* 1:3–48. (Trans. in part E. Fischer, Geographical science in Germany during the period 1933–45; a critique and justification. *Annals of the Association of American Geographers* 39:99–137.)

Volz, W. 1926. Der Begriff des "Rythmus" in der Geographie (the concept of "rhythm" in geography). *Mittelungen des Gesellschaft für Erdkunde zu Leipzig*, 8–41.

————. 1932. Geographische Ganzheitlichkeit (The geographic whole). *Berichte der Sächsischen Akademie der Wissenschaften, Math.-phys. Klasse* 84:3–26.

Wardenga, U. 1987. Probleme der Länderkunde? Bemerkungen zum Verhältnis von Forschung und Lehre in Alfred Hettners Konzept der Geographie (Problems of regional geography? Remarks on the relationship of research and teaching in Alfred Hettner's concept of geography). *Geographische Zeitschrift* 75:195–207.

————. 1989 (forthcoming). Ernst Plewe 1907–1986.

Wittfogel, K. 1929. Geopolitik, geographischer Materialismus und Marxismus. *Unter dem Banner des Marxismus* 3:17–51, 485–522, 698–735.

# Hartshorne, Hettner, and
# *The Nature of Geography*

KARL W. BUTZER

Department of Geography, University of Texas at Austin, Austin,
TX 78712

Richard Hartshorne was one of the foremost advocates and arbiters of methodology in American geography during the quarter century following publication of *The Nature* (1939). Like Carl Sauer, his chief philosophical adversary and rival, Hartshorne appears to have drawn much of his intellectual inspiration from European geography, primarily from German-language literature. Both men selected from or emphasized different authors or works, to reach somewhat different conclusions, and in this they unwittingly complemented each other, in explicating the traditional roots of American geography in nineteenth-century German geography. Finally, Hartshorne and Sauer primarily emphasized these Old World roots in their first methodological statements (Sauer 1925; Hartshorne 1939). Subsequent elaborations of their positions (especially Sauer 1941; Hartshorne 1959) represented more mature, original, and personal statements that reflected what had become two strong American traditions of geography.

During the thirty years preceding the theoretical revolution of the 1960s, American academic geography was increasingly polarized into what has been described as two traditions or "cultures," a Midwestern (or Eastern) and a Californian (or Western) (Porter 1978, 1988). The Midwestern and East Coast tradition was the older and had evolved at institutions including Chicago, Wisconsin, Clark, and Michigan by the 1920s (see Blouet 1981). It was empirical in orientation and developed from a mix of self-conscious decisions as to academic content and practical field experience in regional analyses. The equally influential California variant emerged after Sauer, a disaffected Midwest regionalist, established himself at Berkeley. Beginning with *Morphology of Landscape*, Sauer (1925) grafted new stalks onto the older root: (a) German and French ideas of cultural-historical research, (b) the powerful theme of Marsh (1864) in regard to human agency changing the face of the earth, and (c) interdisciplinary contacts with American anthropology. As the two traditions diverged, the Midwestern variant became increasingly Anglo-American and contemporary in focus, the California counterpart aboriginal and historical.

*The Nature* comprises two main strands. One of these was Hartshorne's personal encounter with and assessment of the vigorous debate he found underway in Germany between proponents of what he conceived of as geographers espousing either systematic or regional goals. The second was his interpretation of the thrust of research and teaching in Midwestern geography, on his academic home ground. An argument could be made (and equally well challenged) that Hartshorne sought to reconcile what he saw as the best of the German school with what he interpreted as the spirit of the Midwestern tradition (as reflected, for example, in the selections of James and Jones 1954). His goal was to give geography an orderly and unified structure, in which, following Hettner (1927), normative concerns played a larger role than is generally realized (see also Sack, this volume).

With the publication of Sauer's presidential address (Sauer 1941), the rift between the two traditions was complete. As their leading proponents, Hartshorne and Sauer each had major impact on the development of geographical thinking, graduate level instruction, and research formulation—but, by and large, at different institutions in the United States and Canada. Classifying the philosophical flavor of a particular program a generation or more ago would be difficult, if not unproductive, since most larger departments included faculty viewpoints differing from those of the majority, and positions changed over time. It would also be incorrect to assume that the hiring of, for example, one or two Ph.D.s from Berkeley would change the tenor of a department already grounded in the "Midwestern" tradition, quite apart from the fallacious assumption that all Berkeley students shared (or would wish to champion) a uniform outlook on geography (see Spencer 1976). But more often than not, either Hartshorne or Sauer would be read in a particular program, the one to the exclusion of the other.[1] This was the case until the 1960s, when the theoretical revolution moved to center stage in methodological controversies.

# Hartshorne's Interpretation of German-Language Geography

Geography was well established in German and Austrian universities by the end of the last century (see Elkins, this volume). The basic roots of the new academic discipline led back to two very different scholars, Humboldt (see Beck 1959–61) and Ritter (see Beck 1979), the former primarily analytical, the latter strictly synthetic. This latent division between "systematic" and "regional" research and presentation was subsequently compounded when physical and human geography emerged as vigorous subfields. Geomorphology had crystallized as a strong subfield with the texts of Richthofen (1886) and A. Penck (1894). The human aspects of the discipline proved more difficult to handle. Ratzel's (1882–91) *Anthropo-Geographie* marked one such attempt, with his second volume, in particular, proposing a mixed culture-historical and economic methodology. The theme of culture history was elaborated by E. Hahn (1896), who had a major impact on Sauer. Meitzen (1895) laid out the foundation for settlement geography, but a standardized approach, drawn in part from Ratzel and

Meitzen, took much longer to crystallize. Primarily through the incremental efforts of Schlüter (1900, 1906, 1920), the "environmentalist" short-circuit was avoided and emphasis instead placed on the analysis and characterization of cultural "landscapes" and their culture-historical interpretation.[2] By the 1920s cultural geography was a strong and clearly delineated subfield, although still less prestigious than physical geography.[3]

While the development of a uniquely cultural methodology did lead to friction within the discipline, the vitriolic controversy that Hartshorne experienced during his stays in Germany between the world wars was not between physical and human or cultural geography. It concerned the integration of analytical research within a larger synthetic framework, with almost all geographers in agreement as to some form of "regional" context, but much dissent as to the exact means of achieving such a unifying paradigm. Squarely at the center of these cross-currents of the 1920s and early 1930s was Alfred Hettner (1859–1941), a versatile and feisty essayist. Hartshorne was particularly impressed by Hettner, whose well-reasoned and historically grounded positions he respected. Hartshorne rarely dissented explicitly with Hettner's views, giving the erroneous impression that the two were in essential agreement.

Something not transparent from Hartshorne's dissection of this controversy is that its key participants, such as Hettner and Passarge,[4] were physical geographers by origin and process-oriented at that. Hettner produced a "genetic" geomorphology text (Hettner 1921) as well as a book on "dynamic" climatology (Hettner 1930), and there can be no question that he was unwilling to discard "genesis" in his ideal framework of regional integration. For Hettner (1927, 209) "causality is a postulate of scientific research" (all translations mine). He argued that "complete scientific understanding always is only possible when phenomena are not only described but also explained" (Hettner 1927, 135); his views of the content, research, conceptualization, and goals of geography are systematically explicated in genetic terms. He clearly gave precedence to a genetic basis for classification, over one based on logical structure (Hettner 1927, 308).

One searches in vain in Hettner's writings for comments that support the widespread impression of Hettner the arch-regionalist[5]; his texts on geomorphology and climatology, as well as the major emphasis on systematic (*allgemeine*) geography in his key methodological study of 1927 argue to the contrary. The theme of chorology is generally introduced as a conceptual framework at the beginning of his book or its chapters (Hettner 1921, 1927), normative and classificatory approaches are employed as heuristic devices (Hettner 1927), maps and good verbal descriptions are considered as effective tools (Hettner 1927), and regional geography is considered as more central to university teaching than systematic geography (Hettner 1927). But I see no preoccupation with regional synthesis and classification. Although Hartshorne certainly did not represent Hettner as exclusively a regionalist, his selective focus on Hettner's somewhat abstract ideas about chorology[6] probably contributed to this impression.

My personal experiences during four years of doctoral and post-doctoral study at the University of Bonn (1955–59) provide further perspectives.[7] Geographers at the time were divided into two schools. The proponents of one emphasized geomorphology or biogeography in their research, but regularly offered regional courses and exhibited their breadth of interest in culture-historical interpretation during the departmental field excursions central to all geography curricula. Members of the other group studied regional land use or settlement, in Germany or abroad, but offered similar ranges of courses and field trips. I experienced both Troll and Lautensach, leading protagonists of these two approaches, in the field, and found their appreciation of both physical and cultural "landscapes" equally sophisticated.

German geography was always field-oriented, and a prerequisite for a university appointment was a major field study in a foreign country (see also Hettner 1927). This fact does not emerge from Hartshorne's analysis, the bibliography of which is largely devoted to matters of theory. The great regionalist debate was essentially concluded by 1933 (Troll 1947), and during the 1950s, methodological writings were considered increasingly superfluous;[8] students were only advised to read thematic textbooks or papers on processual research, including international items. World regional texts were primarily used in the equivalents of grades 11 to 13, i.e., prior to the university level.[9] Perhaps most important of all is that academic geographers were expected to be competent in both physical and human geography.[10]

In sum, I would argue that in Central Europe by the 1930s: (1) geography was widely regarded as a unified discipline, in which physical and human geography were separate but not polarized; (2) regional integration was regarded as the ultimate goal of geographical synthesis, but primarily for heuristic purposes; and (3) there was little consensus about how best to achieve regional integration, except that time and process were critical components, rather than classification or description.[11] Overall, methodological differences were more a matter of style than of competing paradigms. In any case, the differences were less fundamental than those between the Midwestern and California traditions, with German geography more closely approximating that espoused in Berkeley.

By focusing on the methodological debate and neglecting the field component and study curriculum, Hartshorne inadvertently gave the impression of far more basic differences within German-language geography. But there was no compelling reason why Hartshorne *should* have developed a more representative picture of geography in Central Europe. It is often assumed that *The Nature* was largely designed as an authoritative account of German geography, from which Hartshorne derived a methodology to "impose" upon American geography.[12] This incorrect assumption can probably be attributed to Hartshorne's marked reticence in airing his several substantial differences with Hettner's positions (see also below), perhaps suggesting the erroneous conclusion that Hartshorne accepted Hettner as a Supreme Pontiff.

Hartshorne explicitly sought to identify the field of geography "as it has been produced," to determine its "logical position" among the sciences, to examine methodological proposals offered, to present his own precepts for "a broad but unified field," and to relate the discipline, thus characterized, to its cognate fields (31–32). He aimed to provide "an objective presentation of the past" (34), and *The Nature* shows that this generally included both American and European precedents. He was concerned with achieving the greatest degree of consensus possible (32), by "extensive comparison with the findings of other students" (Hartshorne 1948, 497). His contribution to the understanding of German-language geography was a by-product of this investigation, not a central goal, and he is explicit about limiting that discussion to methodological ideas (32). These caveats are important to the following discussion.

# On Physical Geography

To illustrate his view of how the components of geography should fit together as a whole, Hartshorne presented a visual model (147, fig. 1), showing the various subfields of systematic (analytical) geography as derived from their cognate disciplines, on the one hand, and contributing to and integrated within regional geography, on the other.

Hartshorne was explicit that physical geography was an essential part of the discipline:

> the marked differences in the natural environment of different parts of the world, and the partial dependence of most cultural features on the natural environment, is adequate demonstration of the axiom that physical geography is of fundamental importance in geography as a whole (399).

Nonetheless, he preferred to limit it to a supporting role. Barrows's (1923) call to remake geography as a social science was rejected as an "elimination of the study of physical geography in itself" (123). At the same time, Hartshorne insisted that physical geography was justified only as a descriptive aid in characterizing human environments:

> "A study of the adjustments of man to the environment requires a knowledge of the environment, but this knowledge is logically subordinate, not to be studied for its own sake" (123).

He concurs with the Belgian Michotte (1921) that,

> "If geomorphology is primarily concerned with objects to be studied in themselves ... the point of view is that of a systematic science, in contradiction to that of geography, as a chorographic science" (423–24).

The complaint that a non-genetic analysis of landforms would be a sterile exercise appears to be received with some sympathy (224), but was later dismissed as myopic (Hartshorne 1959, 87–88). He notes that German and American (Midwestern) geography differed significantly in regard to geomorphology, but considers the inclusion of "genetic" geomorphology in German geography to have more historical than logical justification.[13]

Hettner took a quite different position. He believed "the progression from description (*Morphographie*) to interpretation (of landforms) (*Morphologie*) was one of the most important advances of geography" (Hettner 1927, 135), arguing that geography must practice (geo)*morphology* in order to resolve the description of landforms properly (Hettner 1923, 44). It was obligatory to learn the methods of one's cognate discipline, unless that research was carried out jointly with a scientist from that field (Hettner 1927, 203). Observation of modern geomorphic processes in the field was essential (Hettner 1927, 204). In terms of understanding structural landforms, it was appropriate to begin the historical dimension of geomorphology with the mid-Tertiary, but the focus was to be on the development of forms and characteristics of the contemporary land surface ("proximate" causation) (Hettner 1927, 132, 137). Nonetheless Hettner was concerned that some geomorphologists (especially Davisian ones) had neglected the study of landforms and their distribution, in favor of earth history as such, in which case their brand of geomorphology functioned as an independent discipline, the role of which was limited to an auxiliary science for geography (Hettner 1921, 3–4; 1923, 42; 1927, 137–38). But there is no ambiguity that he considered genetic landforms as indispensable criteria for geography; in criticizing a particular aesthetic synthesis of world environments, he faulted the lack of distinction made between mountains with different structural histories, or those once glaciated and those not (Hettner 1933).

Hartshorne ignores all this thinking and instead takes issue with only one oblique reference to genetic attributes (Hettner 1927, 223), arguing that such characteristics lend themselves less well to classification than do functional and non-historical criteria (388–89, 418):[14]

"Phenomena are significant in terms of their relations to other present phenomena of geographic significance rather than in terms of their origins" (391).

Hartshorne implicitly preferred to see "genetic" work limited to concordances between physical phenomena, such as climate and vegetation or soils, or soils and landforms. Although

"It is not the function of the geographer to explain the distribution of any phenomenon . . . he may be concerned with such an explanation in order to interpret the relations of that phenomenon to other geographic phenomena" (418).[15]

Nonetheless a more comprehensive recommendation by Hartshorne seems to relegate physical geography to the role of a tool in devising a system of world subdivision:

"*Ideally*, systematic geography receives from the other sciences, or from general statistical sources, the necessary data concerning the distribution of any phenomenon; it classifies the various forms of that phenomenon in any way that is suitable for geographic purposes—i.e., in terms of characteristics significant to regional character—whether or not such classification is available from other sciences. Further, ideally, it receives from the systematic sciences the explanation of the distribution of the phenomenon, that is, its genesis. Whether it be landforms . . . or political states, the principles of development and the causes of distribution, as such, are the concern of the appropriate systematic fields. Geography starts with those facts and principles—assuming always, of course, that the systematic sciences concerned have provided them—as frankly borrowed material" (424–25, emphasis in the original).

In abstract discussions of highly general principles, there inevitably is much opportunity for misunderstanding, unless a concrete example is considered. For several decades, the introductory text of Finch et al. (1960, and in successive editions since 1936) had served as an "in-house" book at the University of Wisconsin. Despite the influence of Sauer's and Leighly's lecture notes on the first edition (Martin 1988), one may presume that the content and approach of its successive editions increasingly reflected the spirit of the faculty and the program of a department in which Hartshorne was highly respected.[16] Form or pattern as well as classification are obviously emphasized, but process and form/pattern in physical geography are treated at length and with an acceptable level of rigor. If such an approach to physical geography is what Hartshorne had in mind, and I suspect this is not far from the truth, it would mitigate my impressions of a severely restrictive position. Furthermore, Hartshorne's position was explicitly idealistic. As he subsequently clarified, the purpose of methodological statements:

> "is to provide orientation—to recognize a central core within and around which geographers work, radiating out in diverse directions, though ever conscious of where they are" (Hartshorne 1948, 501).

# On Cultural Geography

Turning now to cultural geography, perhaps the key question to ask is what *The Nature* considered appropriate subject matter. The initial, exploratory discussion identifies population numbers and densities; physical characteristics of people; clothing, shelter, artifacts, domesticates, and diet; "the manner and substance of thought, speech, and writing"; "the way in which people eat, dance, walk, or ride"; and their impact on the environment (332–33; compare Hettner 1927, 150, for a similar statement). But eventually this broad array of themes is simplified into population and economic activities, such as agricultural "element complexes" (335–41).

Strong difference is taken with the position that cultural geography should be limited to "visible" or material features (ch. 7) and he explicitly takes Sauer (1927) to task for advocating that cultural (as opposed to human) geography direct its attention to material culture (p. 190). Instead, he argues that:

> "Our thoughts are expressed not only in the things which we make but also in the manner and content of what we speak, sing, or dance, in what we write, vote, or tell the census collector. . . . The student of cultural geography who permits an arbitrary rule to forbid him from studying these observable expressions of culture is depriving himself of the possibility of achieving . . . the interpretation of cultural geographic regions" (235).[17]

Further,

> "The test of landscape representation is but one test, and by no means the most
> reliable test, of the differential character of culture of different areas" (232).

But no concrete methods are suggested as to how to achieve such goals. Schlüter
(1920), who did eventually consider a sophisticated array of variables, is criticized for a priori exclusion of languages, cultures, or states (213). Passarge (1933,
79–83), who proposed a hierarchical structure to integrate material and non-
material traits into a comprehensive whole, is countered with statements that
his concept is "scientifically dangerous" as well as "an unsatisfactory hybrid"
(208).

Although "analysis and classification of cultural phenomena" is one goal
(331), the closer examination of function is encouraged. But caution is urged:

> "It might be well for geographers in general not to plunge headlong into intensive
> studies of any one or two phenomena without first considering their importance,
> both in themselves and in relation to other cultural phenomena that are geograph-
> ically significant" (332).

Historical components are considered appropriate insofar as necessary to understand contemporary patterns, with some empathy for "time-slice" historical
reconstructions (187–88, 224), prompting Sauer's (1941) vigorous rejoinder in
defense of a more versatile historical geography. A later comment underscores
that their differences were far more than a matter of semantics: "The purpose
of such dips into the past is not to trace developments or seek origins but to
facilitate comprehension of the present" (Hartshorne 1959, 106). Strikingly absent is reference to specific cultural processes, such as the domestication, diffusion, and landscape change so dear to the Berkeley school. Instead, with
reference to Leighly (1937), culture history is seen as the "logical extreme of
the historical point of view" and "the antithesis of geography" (Hartshorne
1939, 179). These are uncharacteristically strong statements and one can understand that Sauer was irritated. For Hartshorne, unlike Sauer, geography is
first and foremost a synchronic discipline.

Comparing Hettner on cultural and historical geography, we note similar
concerns to Hartshorne's that a historical approach can become all too historical,
and that geographers should limit themselves to historical explanation of contemporary phenomena or to reconstructions of particular past periods (Hettner
1927, 131–32, 151). But these reservations are interpreted rather more liberally,
to include settlement processes such as Medieval internal colonization, defor-
estation, drainage or irrigation, while the importance of past migrations for
cultural complexity is asserted (Hettner 1927, 146, 290). Altogether, domestication, migration, and changing human impacts are given considerable attention
(Hettner 1929): "culture can only be understood in evolutionary-historical terms"
(Hettner 1927, 271). He recognized the difference between physical evolution,
in early prehistoric times, and later cultural evolution, and offered a debatable
but interesting, biogeographical interpretation of prehistoric racial differentiation (Hettner 1927, 270, 289). He articulated the concept of environmental ad-

aptation (*Anpassung*) (Hettner 1927, 145).[18] He also offered sophisticated suggestions as to collective actions and the conflicting goals of humankind, anticipated the notion of suboptimal decision-making, and recognized different attitudes to nature in different cultures (Hettner 1927, 210–11). It would appear
that Hettner had a more comprehensive and explicit conception of cultural
complexity and its interpretation than did Hartshorne.[19]

A final comment is warranted on Hartshorne's discomfort about how to utilize
perception and aesthetics in regional characterization (215, 218–19): "In any
empirical science of geography there is little need for any of the concepts of
'landscape as sensations' " (168). The German concept of "landscape," particularly as *Kulturlandschaft*, perplexed him by its ambiguity, as more than an area
of a certain relative size, namely as an object with "sensory" qualities (150, 252).
Hettner (1927, 151–55, 211–14, 317–21; 1933), repeatedly and at length, dwelled
on what are now called humanistic concerns. Not surprisingly, he approved of
an aesthetic (and psychological) dimension in dealing with "landscapes," and
what Crowe (1938) labeled their mysticism has been elegantly expanded upon
by Wagner (1974) as their symbolic value. In modern terms, *Landschaft* combined
elements of space and place, a basic sentiment in recent humanistic research
(see Butzer 1978).

In total, Hartshorne's expressed views on cultural geography, and its historical
or humanistic components, seem broad but vague at one level, and restrictive
at another. If the presentation of the human and cultural elements of geography
by Finch et al. (1960) may serve, with due qualifications, as a guide,[20] its somewhat bleak repertoire of information on population and economic activities is
disappointing.

## On Dualism and Integration

Dualism, as used by Hartshorne (70–71, 89), is the dichotomy of physical and
human geography, commonly compounded by the additional distinction between regional (synthetic) and systematic (analytical) geography. Hartshorne
explicitly defended the complementarity of systematic and regional geography,
as favored by German methodologists during the 1920s (456–59, 478). In regard
to the first dichotomy, he seems more evasive, in view of the environmentalist
implications of interrelating physical and human phenomena (120–26). Partly
for this reason, he emphasizes that different laws apply to the physical and
social sciences, yet expresses skepticism of the view that geography could serve
as a bridge between the natural and the social sciences (369), emphasizing
instead the essential unity of science: "The separation of things natural from
things human is possible only in theory, in reality they are interwoven" (368).[21]

These points are amplified in Hartshorne (1959), where, in regard to systematic
studies of physical geography, he argues that they:

"were lacking in coherence and divorced from the full context of reality;[22] in consequence they had only limited appeal to the general student.[23] At the same time

the study of the human aspects of geography, in large part divorced from the physical earth features with which they are in reality interwoven, lost both scientific standing and student interest" (Hartshorne 1959, 79–80).

Physical and human geography do not constitute a duality, simply because they have no individual legitimacy, on philosophical and didactic grounds. Hettner (1927, 126–27) also denied this duality but gives additional different reasons, namely that the chorological study of "nature and man" was harmonious and did not introduce "differences of concept and contradiction" (*Verschiedenheit der Auffassung und Zwiespaltigkeit*).

It is also in a very restricted sense that systematic ("topical") and regional geography were, in 1959, considered as complementary methods of analysis. Whatever their intrinsic validity, it rests upon their usefulness to deal with differences of scale:

"There is no dichotomy or dualism, but rather a gradational range along a continuum from those which analyze the most elementary complexes in areal variation over the world to those which analyze the most complex integrations in areal variation within small areas. The former we may appropriately call 'topical' studies, the latter 'regional' studies, provided we remember that every truly geographic study involves the use of both the topical and the regional approach" (Hartshorne 1959, 121–22, also 143–45).[24]

The distinctive purpose of geography is:

"to observe and analyze earth features composed of the interrelations of diverse elements with each other . . . in studying the interrelation of earth features, geography analyzes those features to the extent necessary to explain their interrelations" (Hartshorne 1959, 80).

Hartshorne introduces the idiographic-nomothetic complementarity, but his discussion of "laws" (378–79) leads to the proposal of "generic" concepts and principles that exclude genesis (386–91) in seeking to explain the relations between variable elements in an area. One searches in vain for anticipation of an "ecological" or systemic conception of interrelationships, as opposed to the integration of phenomena in space, convergent at variable scales (Hartshorne 1959, 144).

Although the differences between Hartshorne and Hettner on integration and generalization are subtle, I feel they are noteworthy. Hettner, after reviewing the different subfields of geography, discusses the comparative value and practicality of inductive versus deductive approaches, concluding that geography is best served by a combination of deductive and comparative approaches (Hettner 1927, 185–95). He recognizes the value of particularistic (idiographic) information, but insists that geography, as a science, must also aspire to normative (nomothetic) generalizations, specifically a body of general "laws," which he considers indispensable for aggregate human behavior as well (Hettner 1927, 221–24, 266–73). The emphasis in classification is on genetic characterization and cause-and-effect relationships, not location and description (Hettner (1927, 275–317), and he uses what was then a sophisticated concept, namely "causal chains" (*Ursachenreihen*) (Hettner 1927, 273–75), to interrelate multiple factors in causality. In general terms, he articulated some basic premises of theoretical revolution, and he hints at a systems approach.

# Conclusions

*The Nature* presents a fairly narrow conception of geography. It casts the geographer less as a primary scholar than as a synthesizer, heavily dependent in matters of interpretation on the results of cognate disciplines. Although Hartshorne (1948, 501) may have approved of interdisciplinary "cross-fertiliza-tion" in principle, he seems overly concerned with defining boundaries for the field of geography in an era when the social sciences were fragmenting and when, at least in the U.S., the claim of geography to the physical-environmental arena was actively challenged by geology and meteorology. Not surprisingly, perhaps, the spirit of his construction weighed heavily against such contacts or in-depth research. But while he explicitly considered it important to raise the question "What is geography?", he found it reprehensible to ask "Is it geog-raphy?" of any specific study by a geographer (Hartshorne 1948, 497).[25] He was an idealist and saw methodology as a matter of dialogue, with no individual or group of persons having the authority to define orthodoxy (Hartshorne 1948, 498–99). Unfortunately, this did not hinder some of his disciples from attempt-ing to do exactly that. During my time at the University of Wisconsin (1959–66), doctoral dissertations regularly had to be defended as geographical in con-tent and method. One can therefore legitimately wonder about the inadvertent negative impacts of *The Nature* on the vitality of human and, especially, physical geography.[26]

Contrary to a widespread impression, Hartshorne did not attempt to represent geography as it was implemented in research and teaching in the German-speaking countries of Europe. He explicitly limited himself to the methodolog-ical writings. But even here there is a problem, because Hettner, whom Hart-shorne held in especially high regard, believed in the precedence of (scientific) causality over (epistemological) logic (Hettner 1927, 280). Unlike Hartshorne, Hettner advocated the fundamental importance of physical geography, the "ge-netic" component of geomorphology, considered culture-historical processes pertinent to understanding cultural phenomena, and unequivocally admitted the propriety of humanistic interests such as "aesthetics." Hettner was not an arch-regionalist; he advocated regional characterization but opposed the concept of geography as the study of distributions and held firm views on the merits of "systematic" geography that are exemplified by textbooks in geomorphology and climatology. This fundamental divergence of Hettner and Hartshorne is greatly underappreciated in the literature, and it will come as a total surprise to many that Hettner's position was closer to Sauer's than to Hartshorne's.

Without a parallel analysis of Hartshorne and Hettner, it has simply been assumed that differences were minor, because Hartshorne never criticizes Hettner and very rarely says he disagrees with him. One might be tempted to infer that Hartshorne preferred not to differ in public with a man he revered as the old master. More to the point, perhaps, is that Hartshorne found himself in agreement with Hettner on so many aspects that the outstanding differences, at least to him, seemed unimportant (see note 102, 418).

*The Nature* advocated a regional integration of the discipline in which contemporary, economic elements were presented in a context of applied environmental science. In Germany, Austria, and Switzerland, the emphasis was tilted toward biophysical and cultural components, with much greater latitude for "ecological" interpretation of patterns, and historical analysis of human landscape modification. Just how fundamental this different conception of comprehensive regional integration was can be inferred from the fact that geographers in Central Europe were expected to be competent in both physical and cultural geography, and that "landscape" analysis was diachronic in spirit. On the whole, systematic geography was considerably stronger than it was in the Midwestern and East Coast traditions.

This places *The Nature* in a very different context, as representing Hartshorne's own views and as a uniquely American statement. An appropriate question to explore is the degree to which *The Nature* represented Hartshorne's conception of the methods and rationale of Midwestern geography, perhaps even a consensus emerging during the 1930s among Midwestern geographers.[27]

During the last thirty years, the disciplinary practice of American geography has changed dramatically, first with the theoretical revolution, and subsequently with the vigorous growth of new subfields, usually with strong interdisciplinary ties (see Gaile and Willmott 1989). This demonstrates an old point that no theory of science can hope to delimit a field or constrain its periodic expansion or shifts of direction. Of course, we are no closer to a coherent and practicable, unifying paradigm than we were in 1939—nor, for that matter, are any of the other sciences, social or otherwise. The premises and assumptions have, however, changed, and with them the logical superstructure. In its conservative tone and narrow construction, *The Nature* did not and could not anticipate the productive diversification of our enterprise. It is ironic that some of the many heterodox ideas current in Germany between the world wars, either bore good fruit in American geography or anticipated other, recent intellectual developments. I am thinking here of humanistic perspectives on "place" and the realm of aesthetics, of Hettner's call for a body of scientific "laws," of normative approaches to human geography such as central place theory, and of "ecological" integration via multiple interrelationships, as ultimately facilitated by general systems theory.

## Notes

1. At McGill University in 1954, I was required to read Sauer but not Hartshorne. Needless to say, I remedied this after receiving my first appointment at the University of Wisconsin in 1959.

2. Schlüter's methodology includes what would now be described as demographic and economic patterns and processes, as well as "morphological traits" (the latter devised in analogy to "features" in physical geography, see Bartels [1968, 131]). Sauer's approach is somewhat similar, but distinct, and he first utilizes Schlüter in 1927. The author most often referred to by Sauer (1925) is S. Passarge; other repeated citations include Hettner, A. Penck, Humboldt, N. Krebs, K. Sapper, the historian O. Spengler, the French P. Vidal de la Blache, and the Swedish S. De Geer. Sauer (1927) emphasizes E. Hahn, A. Meitzen, and F. Ratzel. But perhaps most influential in Sauer's own research, apart from Hahn, were R. Gradmann's precedent of reconstructing a pre-cultural "natural landscape" and the translation of J. Brunhes's text on French human geography. The shifting bibliographic emphases in Sauer (1925, 1927, 1941) represent evolving but logical positions that are a little difficult to characterize with precision.

3. Cultural geography in Central Europe was a part of human geography (*Geographie des Menschen*) which embraced the following formal subdivisions: settlement geography and the human modification of the earth, population, transportation, military geography, and economic geography (including *genre de vie* and non-Western cultures); historical and aesthetic perspectives were variably considered as special directions or as formal subfields (Hettner 1927, 142–51). What probably would be considered as cultural geography today therefore focused on settlement, socioeconomic phenomena, culture-historical perspectives and the human use of the earth, as well as aesthetics, at least to some degree.

4. Passarge (1933) is perhaps the most useful example, preceded by three earlier volumes on the same basic theme (*Landschaftskunde*). These works seem a little half-baked, despite some very good ideas.

5. This stereotypic view was current among graduate students at Wisconsin in the early 1960s, and I have heard it offered verbally at meetings of the AAG as recently as 1980.

6. For a more evolved presentation of chorology and the regional approach, one must turn to Lautensach (1952), who devised an elegant model of gradational geographic change across a three-dimensional earth surface. Not cited by Hartshorne is Philippson (1904), on the Mediterranean world, perhaps one of the most successful prototypes of regional-historical analysis and synthesis, influencing other regional studies written in German for more than half a century.

7. Many of the faculty and outside speakers at the regular institute colloquia or the meetings of the Bonner Geographische Gesellschaft had already had teaching appointments or were trained during the 1920s and 1930s. It was therefore possible to compare the literature from between the two world wars with views expressed on methodology as well as reminiscences of the great debate and the personalities of its participants.

8. I recall a long conversation with Hans Bobek on geographical methodology in 1965. When we concluded, he thanked me for my concern with the subject, noting sadly that no one seemed interested any longer. A similar trend was apparent in the United States by that time, where the theoretical revolution was in full swing.

9. This was the case during the 1950s, but less so during the 1930s (see Hettner 1927, 456–59; 1931). But even before World War I, simpler *Allgemeine Länderkunden* were standard fare for the pre-university "matriculation," as I can verify from used books in my collection.

10. Hartshorne acknowledges this indirectly: "The geographer in Germany is, by his training, a geomorphologist as well" (423). Hettner (1923, 44; 1927, 268) even recommends an initial training in geomorphology for human geographers as good for personal scientific development.

11. World maps of Koeppen climates or other classifications were occasionally displayed in the classes I attended in Bonn, but they were only referred to for didactic purposes. By contrast, graduate students at the University of Wisconsin as late as 1962 were expected

to memorize long lists of salient characteristics of the various Koeppen "types." Hettner (1927, 284–85) lauds Koeppen for emphasizing the impact of climate for hydrology and vegetation in his classification, but faults the lack of genetic criteria in devising it.

12. As Hartshorne (1948) rightly observed, far more methodological discussion takes place orally, and casually, than in publication.

13. A bit gratuitously, Hartshorne suggests that "many" geographer-geomorphologists have introduced confusion into methodological thought (424).

14. It needs to be clarified that "time and genesis" meant something very different for geomorphologists in 1930 than it does today. As a student, Hartshorne was exposed to Davisian concepts, including "loaded" genetic terms such as monadnocks or resequent streams. While providing minimal information on process, form or function, such terms claimed to identify a particular evolutionary stage in a time-dependent (but undated), deductive model of landscape evolution, possibly spanning hundreds of millions of years. Geologists *sensu strictu* have never had any patience for relative as opposed to stratigraphic or absolute time. More recent geomorphologists would be primarily interested in Mt. Monadnock as an erosional remnant modified by glacial processes. In Germany, Hartshorne encountered two types of geomorphology, although he may not have distinguished them. One of these, being phased out during the 1930s, focused on tectonic geomorphology and erosional mega-landscapes, dating back to the Tertiary or even the Paleozoic. Hettner, who considered the Davisian approach as a "premature, purely hypothetical" substitute for causality (Hettner 1923, 44), emphasized structure and process in his own textbook (Hettner 1921). The other kind of geomorphology, derived from Quaternary and "climatic" geomorphology and dominant in Germany since 1940, was concerned with "process and form" in deciphering much more recent landscapes. Although Hartshorne does not appear to have followed the geomorphological literature, one might presume that he would have been more receptive to the latter.

15. "Climatic" geomorphology seems to approximate this idea, in that it accepts lithologic and tectonic features as given, and seeks to understand the impact of climate and climatic change in shaping contemporary landforms via the plant cover, weathering, and soils (see Butzer 1976, ch. 16–20). This approach had its origins in Germany during the 1930s and reflected a growing concern by geomorphologists themselves for a more geographical methodology, even more narrowly set than in Hettner's (1921) textbook. Nonetheless, Hartshorne (1959, 75, also 124) concludes that such phenomena "form only a very weak degree of integration."

16. Use of this book was mandatory for introductory courses, and I was variously cautioned 1960–62 by mid- and high-level faculty to adhere to its contents.

17. The issue was whether phenomena had to be *sinnlich wahrnehmbar* (Schlüter 1906; Hettner 1927, 128–29), i.e., amenable to sensory perception, which is not identical with the narrower concepts "visible" or "material."

18. Hettner was not an environmental determinist, even though he favored trying to relate cultural phenomena to environmental conditions (Hettner 1927, 210). The physical landscape was to be studied first, followed by examination of the cultural counterpart at a generalizing, deductive level. Just how he proposed to interpret the linkage is illustrated by the example that topography influences communication routes and hence the location of many but by no means all settlements. He argued that environmental influences vary according to cultural complexity (Hettner 1927, 211) and that people do not respond identically to the same stimulus, because their particular character and historical experience are paramount. Geography should be primarily concerned with collective actions (patterns of behavior), although the role of individuals may be decisive, and outcomes need not always be rational or predictable (Hettner 1927, 267). Hettner was zealous in his search for causality and injudiciously equated the term "determinism" with causality. But the particular context leaves no doubt that he was arguing about the need to establish normative patterns of human behavior. Wherever one may stand in

regard to the dialectic between the individual and society, it would be inappropriate to categorize Hettner as an environmental determinist. This conclusion is born out by his posthumous general volume on anthropogeography (Hettner 1947).

19. Sauer (1925) cited Hettner (1923) with evident approval and presumably would have been sympathetic with the presentation of culture history in Hettner (1929), although that somewhat obscure treatise is not referred to in later publications of Sauer's.

20. A comparison of these segments of Finch et al. (1960) with Hartshorne and Dicken (1935, 1938) supports such an assessment. On the other hand, I have no evidence that Hartshorne viewed the human geography component of Finch's book as reflective of his views, that he endorsed these views, or that Finch sought to express Hartshorne's views.

21. According to Hettner (1927, 126–27), geography was both a natural science and a humanity, and belonged in both, with the potential to serve as a bridge between them.

22. Primarily so because such studies "attempted to separate man from the rest of nature" (Hartshorne 1959, 79). Fortunately, we now appreciate more clearly, following Glacken (1967), that man is both a part of and apart from nature.

23. This is another unambiguous signal that Hartshorne's view of geography did not represent how it was practiced in Central Europe; the comment is also not in accord with Hettner (1921, 1927, 1930). Whether physical geography bored general students in America is at the very least debatable. In as far as there were fundamental problems in the successful teaching of introductory physical courses during the 1950s and 1960s, I would place the blame on their staffing by beginning instructors, many of whom lacked the professional training to do competent research in physical geography.

24. Hartshorne's note 9 in this place states, "The viewpoint presented in this section . . . differs markedly from that presented in *The Nature of Geography*" in regard to the essential continuity between systematic and regional approaches. But much more is at stake here, whether in fact Hartshorne has abandoned systematic geography altogether. I incline to see the 1959 position as a much narrower construction, but one that may be hinted at in Hartshorne's 1939 characterization of regional geography: "It must first express, by analysis and synthesis, the integration of all interrelated features at individual unit places, and must then express, by analysis and synthesis, the integration of all such unit places within a given area" (467). The words of the 1961 abstract (xv, as added to Hartshorne 1939), that systematic and regional geography are "mutually dependent" approaches, that "must be combined in specific studies," are sufficiently ambiguous to bridge rather than identify the differences. In any case, the change represents a sharp divergence from Hettner (1927), who emphasized rather than questioned the integrity of systematic geography as such.

25. "For practical or personal reasons, students specializing in particular topics of systematic geography will pursue such studies," with the proviso that "this imposes no obligation on other geographers to attain such competence" (Hartshorne 1959, 107). On the basis of Hartshorne's comments during the regular weekly faculty meetings or at doctoral proposal hearings or defenses at the University of Wisconsin, as well as in private conversations, as recently as 1985, I too can vouch that he was sincere and consistent about this, never raising the question "Is it geography?"

26. Depending on the selection of editors for major professional journals or of directors and panels controlling government granting agencies, particular brands of orthodoxy can indeed inhibit research in certain subfields of a discipline. It is a fact that physical geographers were long alienated on both counts. Similarly, the long-term retrenchment of faculty appointments in physical geography proved embarrassing to the discipline when Earth Day 1970 created a groundswell of new ecological interest among students, an opportunity to which geography was unable to respond because of ossified, descriptive curricula and a dearth of qualified researchers. But to ascribe this deep-seated problem to any one factor would be simplistic, particularly when its origins go back to before

1939 and since it only came to a head during the years of the theoretical revolution. One can also not ignore inherent problems within both physical and cultural-historical geography (see, for example, Leighly 1955), both of which acquired renewed vigor with the incorporation of new ideas or methodologies as a direct or indirect result of the theoretical revolution. Nonetheless, the majority opinion in many departments towards "systematic" geography was pejorative for several decades, effectively marginalizing its practitioners and discouraging student interest. It therefore bears examination how this intellectual retrenchment to a narrowly-defined core, suspicious of processual studies or interdisciplinary contacts, came to be characteristic of so many departments by the 1960s. My surmise is that it was a logical development of the Midwestern-East Coast tradition, far more than it was a possible response to Hartshorne's positions in *The Nature*. Almost 250 geographers (including most of the best) were engaged in some form of wartime government service in the Washington, DC area during 1942–45 (see Martin 1988). They were primarily employed in writing regional position papers or in cartography, and it seems inevitable that the regionalist and economic proclivities of the Midwestern-East Coast tradition would have then crystallized into a coherent paradigm, reinforced by close proximity and strong personal ties, that endured for many years as an elite club.

27. A possible starting point is the descriptive land-use mapping project at Montfort, Wisconsin, which served as a model for numerous subsequent studies, that can probably be considered representative of the Midwestern tradition during the 1920s and 1930s (see James and Martin 1978, 73–77). The research activities of geographers in the Office of Strategic Services, the War Department, and the Intelligence Service would appear to be a second, promising avenue of study.

# References

**Barrows, H. H.** 1923. Geography as human ecology. *Annals of the Association of American Geographers* 13:1–14.

**Bartels, D.** 1968. *Zur wissenschaftstheoretischen Grundlegung einer Geographie des Menschen.* Wiesbaden: F. Steiner (Geographische Zeitschrift, Beiheft 19).

**Beck, H.** 1959–1961. *Alexander von Humboldt.* Wiesbaden: F. Steiner, 2 vols.

———. 1979. *Carl Ritter, Genius der Geographie.* Berlin: D. Reimer.

**Blouet, B. W.,** ed. 1981. *The origins of academic geography in the United States.* Hamden, CT: Archon.

**Butzer, K. W.** 1976. *Geomorphology from the earth.* New York: Harper and Row.

———. 1978. Cultural perspectives on geographical space. In *Dimensions of human geography: Essays on some familiar and neglected themes,* ed. Karl W. Butzer, pp. 1–14. Department of Geography Research Paper 186. Chicago: University of Chicago.

**Crowe, P. R.** 1938. On progress in geography. *Scottish Geographical Magazine* 54:1–19.

**Finch, V. C.; Trewartha, G. T.; Robinson, A. H.; and Hammond, E. H.** 1960. *Elements of geography: Physical and cultural.* New York: McGraw-Hill, 4th ed.

**Gaile, G., and Willmott, C. J.,** eds. 1989. *Geography in America.* New York: Merrill.

**Glacken, C. J.** 1967. *Traces on the Rhodian shore: Nature and culture in western thought from ancient times to the end of the eighteenth century.* Berkeley: University of California Press.

**Hahn, E.** 1896. *Die Haustiere und ihre Beziehungen zur Wirtschaft des Menschen.* Leipzig: Duncker und Humblot.

**Hartshorne, R.** 1939. *The nature of geography: A critical survey of current thought in the light of the past.* Annals of the Association of American Geographers 39:173–658 (Reprinted Lancaster, PA: Association of American Geographers, 1946; reprinted with notes and corrections, 1961).

———. 1948. On the mores of methodological discussion in American geography. *Annals of the Association of American Geographers* 38:492–504.

————. 1959. *Perspective on the nature of geography.* Washington: Association of American Geographers.

Hartshorne, R., and Dicken, S. N. 1935. A classification of the agricultural regions of Europe and North America on a uniform statistical basis. *Annals of the Association of American Geographers* 25:99–120.

————, and ————. 1938. *Syllabus in economic geography.* Ann Arbor, MI: Edwards, 2nd ed.

Hettner, A. 1921 (1972). *The surface features of the land: Problems and methods of geomorphology,* rev. ed. 1928, trans. P. Tilley. New York: Hafner.

————. 1923. Methodische Zeit- und Streitfragen. *Geographische Zeitschrift* 29:37–59.

————. 1927. *Die Geographie: ihre Geschichte, ihr Wesen und ihre Methoden.* Breslau: Hirt.

————. 1929. *Der Gang der Kultur über die Erde.* Leipzig: Teubner, 2nd ed.

————. 1930. *Die Klimate der Erde.* Leipzig: Teubner.

————. 1931. Die Geographie als Wissenschaft und als Lehrfach. *Geographischer Anzeiger* 32:107–17.

————. 1933. Zur aesthetischen Landschaftskunde. *Geographische Zeitschrift* 39:93–98.

————. 1947. *Allgemeine Geographie des Menschen: Vol. 1. Die Menschheit,* ed. H. Schmitthenner. Stuttgart: W. Kohlhammer.

James, P. E., and Jones, C. F., eds. 1954. *American geography: Inventory and prospect.* Syracuse, NY: Syracuse University Press.

———— and Martin, G. J. 1978. *The Association of American Geographers: The first seventy-five years 1904–1979.* Washington: Association of American Geographers.

Lautensach, H. 1952. *Der geographische Formenwandel: Studien zur Landschaftssystematik.* University of Bonn, Geographisches Institut, Colloquium Geographicum 3.

Leighly, J. 1937. Some comments on contemporary geographic methods. *Annals of the Association of American Geographers* 27:125–41.

————. 1955. What has happened to physical geography? *Annals of the Association of American Geographers* 45:309–15.

Marsh, G. P. 1864. *Man and nature: Physical geography as modified by human action.* Reprinted and ed. D. Lowenthal, Cambridge, MA: Harvard University Press, 1965.

Martin, G. J. 1988. Preston E. James, 1899–1986. *Annals of the Association of American Geographers* 78:164–75.

Meitzen, A. 1895. *Siedelung und Agrarwesen der Westgermanen und Ostgermanen, der Kelten, Römer, Finnen und Slawen.* Berlin: Hertz, 4 vols.

Michotte, P. 1921. L'orientation nouvelle en géographie. *Bulletin, Société royale belge de Géographie* 45:5–43.

Passarge, S. 1933. *Einführung in die Landschaftskunde.* Leipzig: Teubner.

Penck, A. 1894. *Morphologie der Erdoberfläche.* Stuttgart: Engelhorn, 2 vols.

Philippson, A. 1904. *Das Mittelmeergebiet: Seine geographische und kulturelle Eigenart.* Leipzig: Teubner. 4th ed., 1922.

Porter, P. W. 1978. Geography as human ecology: A decade of progress in a quarter century. *American Behavioral Scientist* 22:15–39.

————. 1988. Sauer, archives, and recollections. *The Professional Geographer* 40:337–39.

Ratzel, F. 1882–1891. *Anthropo-Geographie: 1. Grundzüge der Anwendung der Erdkunde auf die Geschichte; 2. Die geographische Verbreitung des Menschen.* Stuttgart: Engelhorn.

Richthofen, F. von. 1886. *Führer fur Forschungsreisende.* Berlin: Oppenheim.

Sauer, C. O. 1925. The morphology of landscape. *University of California Publications in Geography* 2:19–54 (Reprinted in *Land and life: A selection from the writings of Carl Ortwin Sauer,* ed. J. Leighly, pp. 315–50. Berkeley: University of California Press, 1963).

————.. 1927. Recent developments in cultural geography. In *Recent developments in the social sciences,* ed. E. C. Hayes, pp. 154–212. New York: Lippincott.

————. 1941. Foreword to historical geography. *Annals of the Association of American Geographers* 31:1–24 (Reprinted in *Land and life: A selection from the writings of Carl*

*Ortwin Sauer,* ed. J. Leighly, pp. 351–79. Berkeley: University of California Press, 1963).

———. 1974. The fourth dimension of geography. Annals of the Association of American Geographers 64:189–97.

**Schlüter, O.** 1900. Die Formen der ländlichen Siedelungen. *Geographische Zeitschrift* 6: 248–62.

———. 1906. *Die Ziele der Geographie des Menschen.* Munich: Oldenbourg.

———. 1920. Die Erkunde in ihrem Verhältnis zu den Natur- und Geisteswissenschaften. *Geographischer Anzeiger* 21:145–52, 213–18.

**Spencer, J. E.** 1976. What's in a name? "The Berkeley School." *Historical Geography Newsletter* 6:7–11.

**Troll, C.** 1947. Die geographische Wissenschaft in Deutschland in den Jahren 1933 bis 1945: eine Kritik und Rechtfertigung. *Erdkunde* 1:3–48.

**Wagner, P. L.** 1974. Cultural landscapes and regions: Aspects of communication. *Geoscience and Man* 10:133–42.

# The Nature of Geography: Post Hoc, Ergo Propter Hoc?[1]

FRED LUKERMANN

Department of Geography, University of Minnesota, Minneapolis,
MN 55455

Richard Hartshorne opens *The Nature of Geography* (subtitle—*A Critical Survey of Current Thought in the Light of the Past*) as follows: "For me, therefore, it is important that the work was not planned from beforehand, but grew of itself out of a much smaller idea. This does not mean that it is planless; it has been revamped many times to form an organized whole, but its nature developed from itself rather than from any intention on my part.[2]"

In the 1946 edition of *The Nature* (the second printing), an abstract is provided elaborating on the original Table of Contents and its organization (with additional bibliography, corrections, and supplementary notes). At the 1947 AAG annual meeting, Hartshorne presented a paper "On the Mores of Methodological Discussion in American Geography," (1948) in response to reviews and criticism of *The Nature*; this paper was later appended to the 1961 edition. That edition (reset from the earlier photocopies) was issued with additional corrections after the articles involving the Schaefer paper and Kant (1955 and 1958) and *Perspective on the Nature of Geography* (1959) were published.[3] Given that span of composition (1939–59), I am limiting my comments to published material antedating 1959.

## Antecedents

It serves our purposes to review some relevant antecedents to the 1939 manuscript which throw light on the interest in the discipline in matters concerning the general nature of geography, before discussing, in detail, the Hartshorne tour de force.

The title, *The Nature of Geography*, was used only once before, to my knowledge (other than in textbook headings), as the title of a 1926 review by Charles Redaway Dryer of Carl Sauer's "Morphology of Landscape" in the *Geographical*

*Review*.[4] This is the only formal review of that important essay in the American disciplinary press; it is not cited by Hartshorne in the original 1939 edition or in the final 1961 edition.

Among American geographers in the early years of the present century, there had been an active interest in the nature of the field. Without being exhaustive, one can mention several articles by William Morris Davis (including one on Alfred Hettner) in the first decade of the century in *The Annals*; Dryer also gave a paper on "Philosophical Geography: Strabo, Kant and Bain" in 1907, followed in 1919 by his presidential address, "Genetic Geography: the Development of the Geographic Sense and Concept" (1920). In 1918, Nevin M. Fenneman's presidential address, "The Circumference of Geography" (1919), attracted wide attention, followed by more numerous names in the third and fourth decades of the century, including the aforementioned "Morphology of Landscape" and "Recent Developments in Cultural Geography," published apart from *The Annals* by Sauer in 1925 and 1927.[5] Many, including Sauer, were dependent in their commentaries on Hettner's early papers, but second-hand Kantian and Vidalian themes were also present in the American literature.

The fourth decade of the Association of American Geographers was the premier decade for methodological controversy. Almon Parkins, Isaiah Bowman, Vernor Finch, Glenn Trewartha, Preston James, and Robert Platt, to name a few writers, and numerous others at AAG meeting roundtables were participants. The discussions centering on the methodology of regional description and analysis were prevalent but not necessarily prevailing. Above all, two articles by Leighly in 1937 and 1938 received major attention.[6] The 1937 article, entitled "Some Comments on Contemporary Geographic Methods," especially raised Hartshorne's ire. It is recorded:

> At a luncheon meeting (following a paper by R. S. Platt replying to the Leighly article), a discussion ensued and Hartshorne informed Whittlesey, who was the editor of the Association's *Annals,* that the paper should have been consigned to the circular file. Whittlesey requested that Hartshorne state his criticisms in a form that could be published. There began a study which, with a sojourn in Europe and encouragement from Whittlesey, evolved into *The Nature of Geography*, first published in two issues of the *Annals*[7] (James and Martin 1978, 81).

The controversy with Leighly (both the 1937 paper, 16 pages, and the 1938 paper, 20 pages) is covered in nineteen citations in *the Nature of Geography*, approximately evenly divided between "Some Comments . . ." and "Methodologic Controversy in Nineteenth-Century German Geography."[8]

Parallel discussions in the British geographic literature, especially in 1938–39 between Crowe, Dickinson, Forde and Stevens (stemming in part from German, French and American sources) on progress, landscape, society, region and history in the *Scottish Geographical Magazine* attest to both the importance of methodological discussions in the discipline and the willingness to dispute the *nature* question in each generation of geographers.[9]

# From *The Nature* to *Perspective*

The publication of *The Nature* and the interlude of World War II dampened the flames of controversy for a while. The 490 pages in the *Annals* probably dampened more flames than the war. Only Sauer really took offense in print, in his 1941 presidential article "Foreword to Historical Geography" in the *Annals*.[10] Klimm's review of 1947 documented *The Nature*'s acceptance as *the* text in whatever courses in the History of Geography or the Development of Geographic Thought were offered in the discipline, excepting, of course, Berkeley, where Leighly presided over the curricular offerings.[11]

The impact of *The Nature* on the profession was extensive and dominating, literally shutting out other views and other versions of the history of geographic thought until the early fifties. Because of that dominance, both its pervasiveness and its presumed authoritativeness, when reaction came it was sharp, strident, basically polemical and was perceived by most geographers as radical and revolutionary. Neither was the case.

The Schaefer attack of 1953 was familial, within the Hettner circle, but it was sharp and knowing. It was also flawed. Schaefer failed to complete his manuscript before his death, and a colleague, the philosopher of science Gustav Bergmann, edited the paper before publication. An unfortunate aspect of this circumstance was a lack of extensive citation of references and undoubtedly a selective perspective on several paragraphs concerned with historicism, morphologic laws and emergent holism. These were deep concerns of Bergmann's and reflected not only his and Schaefer's viewpoint but also that of the logical empiricist school of positivist philosophy in the United States and in central Europe, but not necessarily a knowing geographical viewpoint.[12]

Schaefer's paper opens with a firm commitment to a scientific systematic geography approach to research as a base for regional geography. This position is found in Hettner, Sauer and Hartshorne, but not as strongly polarized as in Schaefer. It follows that a common base of sources and authorities were being appealed to—a typical family quarrel appealing to a priori authority and error. For Schaefer, the exceptionalist culprit was Immanuel Kant; for Hartshorne, the authority was Immanuel Kant—through Hettner.[13]

Hartshorne's response to Schaefer was unforgiving, and Schaefer, whether understood or not, became a symbolic martyr for the proactivist wing of younger geographers who invoked "lawfulness," "quantification" and "modeling" in the second wave of post-war geography after 1958.[14] *Perspective on the Nature of Geography*, published in 1959, and the 1961 edition of *The Nature*, although ameliorative and conciliatory from Hartshorne's viewpoint, only alienated the "scientific" wing more. In fact, Schaefer, in his emphasis on morphologic (cross-section) laws and on structural taxonomies such as Christaller's central place studies, was not as process-oriented as the coming generation desired or needed.[15]

Thus, the attack by the "quantifiers" and spatial-interactionists against *The Nature* was equally sharp and strident—and ill-conceived. Hartshorne was depicted as the arch idiographer, protector of the unique, and guardian of the past. Yet, in truth, he was by predilection and in his teaching and research, liberally positivist in his position in the discipline. His attraction to Hettner, another liberal positivist, was predictable; his list of publications in economic and political geography and his earlier location studies spoke to a strong interest and practice in systematic geography. Where Hartshorne differed from Hettner (which should have been congenial to the spatial-interactionists), in his opposition to genetic explanation, for example, somehow in his coolness toward distribution-and-process (cause-and-effect) studies in geography, and contrarily, his predilection for functional (non-causal, teleological) studies, he ran afoul of the trend and direction of the late fifties and sixties.[16]

The other major trend in post-war geography, toward planning and environmental-ecological studies, while systematic in approach, was also normative in outlook and received no particular support in the pages of *The Nature* or *Perspective*, even though Hartshorne may have been personally interested in those fields.

The response by the historical/cultural/landscape groups was muted, yet they were probably the most misunderstood in *The Nature* and they, in a sense, had the most to lose. As noted above, Sauer responded with some personal bitterness in 1941 but refused to enter into any dialectic with Hartshorne or *The Nature* on substantive issues. Sauer found it distasteful and, he claimed, unproductive, to engage in methodological debate; you state your case, then get on with your work—an example, was his preferred response. Whether the chasm between Hartshorne's views and the body of cultural, historical geographers was as wide and deep as they contended is hard to judge. It was said that he just did not understand (being a positivist) that historical geography could not be sloughed off as just "past geographies." The answer must lie in basic assumptions and not in methodological practices. As to landscape, the extensive discussion in *The Nature* over "material, physically observable, features" is a perfect illustration of minds not meeting.[17] Insofar as the ideas of history, culture and landscape are inseparably linked in contemporary geography, they deserve the more detailed discussion given in the sections on *Historical Geography* and *Landscape* below.

A synopsis of the table of contents of *The Nature* reveals why the reaction in the fifties was so deep and bitter. In the abstract to the 1946 second printing, Hartshorne writes that "some readers may wish to omit those sections totalling nearly one-third of the total text that arrive at negative conclusions" (vii).[18] The most critical was Section III, entitled "Deviations from the Course of Historical Development" and covering attempts in the discipline to construct a "scientific" geography, basically recalling the German and English literature as to geography being a science of the planet earth, the science of distributions, or the science of relationships. In Sections V, VII, IX and X, there are discussions of the landscape/*landschaft* confusion, comments on the limitation of phenomenon to sensory data, the region as a concrete unit object, and natural regions and comparative systems of generic regions based on elements of the natural environment.

This litany of "deviations" entered the broader literature in the *Encyclopaedia Britannica*, within Preston James's definitive article on "Geography," in a prominent section entitled "Deviants from the Mainstream."[19] James's article remained effectively unchanged for three decades from the 1950s through the 1970s. The "Mainstream" referred to is the Hettner/Hartshorne version of the *chorologic science* of geography, and therein lay the rub.

From the viewpoint of the disenfranchised—the deviates—who argued for a more systematic, abstract, causal, logical, lawful, processual, nomothetic geography, this was the voice of reaction, the constraint that limited their data collection, their statistical approaches, their theories, their modeling, their laws and their science.

Hartshorne was cast by them as the purveyor of the idiographic past, the mere describer, the peoples and places, the unity and diversity, the differences-and-similarities sing-song siren of the unique, the particular and, above all, as the exceptionalist methodologist against processual and causal explanation. All of this had surfaced by 1959 when *Perspective on the Nature of Geography* was published. The editor wrote in the preface:

> under circumstances described in his Foreword, Professor Hartshorne was moved to make a new statement which would bring up to date his own and others' thinking, and would express a logical concept of geographic investigation and scholarship more directly and in a more affirmative manner than in the earlier work[20] (vii).

In the Foreword, Hartshorne details the impact of *The Nature* on the profession, much as I have done. His apology reads:

> the encyclopedic character of that work [*The Nature*] and the large amount of attention devoted to negative criticism, though essential in both cases to the purpose of the study, tend to obscure its positive conclusions. The criticisms and challenges that have been raised since then, in the literature, in correspondence and in seminar discussions, have demonstrated the need for a reconsideration of ten basic questions each of which form the topic of a chapter in this book[21] (1,2).

The statement obviously implies that "substantial revision" was necessary. Was it successful? Retrospectively, the ten chapters of the *Perspective* form a checklist for an evaluation.

# The Light of the Past

Can the history of a discipline define a discipline for the future? No historian would attempt that. Such a reconstruction would involve a genealogical fallacy if the concepts that compose a discipline at any given period were tethered to an essential, historical source on the basis of a generative or lineage principle. "Generative" in this sense would mean that a geographic idea or concept like "landscape" or "areal differentiation" could be tracked to a source which becomes the basis of its present meaning. Rather, the meaning of a concept at any point must be understood in the context of concurrent historical and cultural factors. An historical survey fleshes out the background of leading ideas and affords insight into their implications and cannot be understood as merely tracking ideas from some independent and specific source.

The generative, essentialist approach distracts from the methodological issues in the air at any given time. It deflects from an interest in landscape or areal differentiation itself—the *contextual* development of the idea—and especially from the interplay of the motives and assumptions which led to the concept in the first place. Whatever the concept, it can be shown to have originated in the past; but the tension, the opposition, which surrounds any idea at any given time gives it its historical meaning and understanding.

As to the above question, Hartshorne states:

> This book follows the principles on which *The Nature of Geography* was based, namely that the determination of the nature, scope and purpose of geography is primarily a problem in empirical research[22] (10–11).

Further:

> This "critical examination in the light of the past" is, of course, not automatic; it involves logical judgment, whether that of the writers quoted or of the author himself. But the logic followed can be tested by the readers at each step; if that is sound, the conclusions follow from the premises. If the reader questions the conclusion, his task is to demonstrate the error in the statement of facts, or in the premises or the logical reasoning[23] (8).

The consequences of such reasoning in *Perspective* resulted in the same basic conclusions as in *The Nature*. The word "deviations" is dropped, but the answers are the same. It is a judgment of the past, not an understanding or an explanation of the past. All in all, as a history of geography, *The Nature* failed if that was its intent. It is a critical examination but not "in the light of the past."

History is defined by Hartshorne by a quote from Alfred Louis Kroeber.

> History is an endeavor at descriptive integration of phenomena in contrast to the method of the systematic sciences which is "essentially a procedure of analysis, of dissolving phenomena in order to convert them into process formulations"[24] (145).

One can only conclude that *The Nature* was not history, but rather as Hartshorne described it in *Perspective*: "it involves logical judgment, whether that of the writers quoted or of the author himself . . . , the conclusions follow from the premises." The "light of the past," in this case, is not a search for history in Kroeber's terms but a search for authority to validate the conclusions drawn from selected premises—largely formulated by Hettner, who had philosophic associations and leanings rather than historian associates.[25]

# Historical Geography

*The Nature* approaches the issue of historical geography from two basic viewpoints, not necessarily related. In attempting to define geography as a discipline, it has been traditional to compare it with history. From that basepoint, a further study of the question could proceed either logically or empirically; Hartshorne chose to take the logical path of a spatial and temporal dichotomy. The origins of the distinction are shadowy, but at least from Varenius onward, the distinction of a chorographical versus a chronological perspective has been definitive in the separation of the two disciplines, but not at the expense of ignoring a methodological parallelism that distinguishes the two disciplines from the other sciences.[26] Insofar as a modern source is cited who can be quoted in detail as to both the parallelism and the dichotomy of history and geography, that authority is inevitably Immanuel Kant—for both Hettner and Hartshorne, that is the authoritative case. From the basic stance of the dichotomy, it can be argued that historical geography logically can be limited to "past geographies"; but that is certainly not an empirical reading of the literature recorded as historical geography.

The Hettner/Hartshorne concept of historical geography as a *total* geography of the past, or more properly past geographies, is actually very unclear when combined with their support of disciplinary classification on their Kantian grounds. In the Einleitung to his *Physische Geographie*, Kant *classified* the sciences as either conceptual or actual.[27] History and geography, according to Kant, were not conceptual; they were concerned with phenomena that were observed in time and space—in actual time/space. Kant was classifying phenomena both as to where they were *found* and where they could be *logically* related. Normally, one would argue, all phenomena could be classified in both ways; but history and geography cannot be split in Kantian terms; time cannot be separated from space in sensory actuality; *conceptually*, Kant did not *classify* history and geography.

Hartshorne's introduction of Kroeber's definition of history undermines his (and Hettner's) interpretation of Kant. In Kroeber's definition, the word "history" can literally be replaced by "geography," or better, "chorology," or Sauer's "landscape," thus emphasizing the parallelism of the four concepts rather than their separation. That fact is exactly what frustrated and muted the voices of the historical, cultural and landscape geographers in responding to the strictures laid down in *The Nature*. For example, in Hartshorne's discussion of Broek's study of the Santa Clara Valley, he posed the insoluble dilemma of how one handles the sequence of cultural landscapes without invoking "history" and "process"—if one accepts his (Hartshorne's) Kantian strictures.[28] Hartshorne's "logic," turned to "judgment," turned off the vast majority of historical and cultural geographers. As Sauer noted (1941,4): "We shall not get very far if we limit ourselves in any way as to human time in our studies"; and more pointedly: "Only if we reach that day when we shall gather to sit far into the night, comparing our findings and discussing all their meanings, shall we have recovered from the pernicious anemia of the but-is-this-geography? state."[29]

Other than that outburst, one can only conclude that *The Nature* and *Perspective* did neither dissuade nor move the historical, cultural geographers from their errant ways. The lack of response to *The Nature* on that score was overwhelming.

## Landscape Geography

Hartshorne's initial encounter with landscape theory in geography was in 1933 in a discussion with Glenn Trewartha over the question of the limitation of the subject matter of geography to "material, physically observable, features." That position had been ascribed to Schlüter and Sauer, and Hartshorne had felt it was a major stumbling block in the development of political geography as a major subfield in the discipline. In his paper on "Recent Developments in Political Geography" in 1935, Hartshorne's differences with Sauer were spelled out in detail. In large measure, Hartshorne's position methodologically, at that time, was dependent on Hettner's philosophical position and criticism of German landscape theory summarized in his 1927 collection of essays from thirty years earlier. This question of observable features had been argued in Germany, ad infinitum, and had been elaborated on in America in Sauer's "Morphology of Landscape" in 1925 and expanded upon in more detail as to contending viewpoints in Sauer's "Recent Developments in Cultural Geography" (1927) and a summary article on cultural geography in the *Encyclopedia of the Social Sciences* (1931).[30]

As is evident to this day, no consensual resolution over the matter was reached, and the issue has separated the historical, cultural, regional geographers as well as the systematic geographers into various and continuingly disputatious camps. As for Hartshorne and Sauer, it was a disagreement that entered into virtually every encounter they had well into the 1970s. The crux of the matter, put briefly, was the failure of Hartshorne to understand that landscape(s) were not observable, they were constructs, they were conceptual; based on the areal *experience* of human beings. Landscapes were *expressions* of the value systems of the culture groups occupying specific areas—regions.[31] Sauer's failure, and the cultural landscape practitioners's failure, was in their refusal to continue the debate after the publication of *The Nature* and *Perspective* and make their viewpoint clear; they basically refused to contend with each other.

Underlying the dispute was the more serious debate of objective science versus subjective values—the empirical versus the experiential. Sauer, as he made evident, could not accept Hartshorne's judgments about landscape or cultural geography. Sauer, as a historical, cultural, landscape geographer, could only be confused by the dialectic in *The Nature*; he could not accept the premises.

## Neo-Kantian Geography

The problem of values haunted geography as it did every other science in the mid-twentieth century. In geography, the issue was conjoined with the late nineteenth-century question of idiographic description versus nomothetic explanation. Neither Hettner, Sauer nor Hartshorne found this Neo-Kantian dichotomy of idiographic/nomothetic particularly relevant to geography, whether Windelband's, Rickert's or Weber's version. They all argued that geography, as a natural and social science, transcended such simplistic classification.

Hettner, who had association and even some intimacy with the Heidelberg-Baden branch of the movement, was not moved by their arguments. The initial question of division between description and explanation, although seemingly relevant to geographic methodological practice, was not particularly applicable to the trends in historical, cultural geography, particularly those geographers who followed Sauer's areal or habitat differentiation thesis. Much has been made of a *Geisteswissenschaften* relevance in recent years to human and humanistic geography, but that is reading back into the period of the "Morphology of Landscape," *The Nature* and *Perspective* much more than the documentary evidence supports.[32]

Schaefer's mention of Rickert and Dilthey is a slim hook on which to hang an epistemological twist to middle twentieth-century American geography, and, in a sense, emphasizes "historicism" and the "unique" where they do not apply. Rickert's variety of Neo-Kantianism more correctly associated the idiographic with an "individualizing" methodology as against "unique," and he preferred the label of *Kulturwissenschaften* in place of history in Windelband's *Geschichtewissenschaften und Naturwissenschaften* label. Neither Hartshorne (nor Sauer before him) speak with any but marginal knowledge of the complex philosophic roots of Neo-Kantianism and reject what they do know of it as to its application in geography. One can make an appealing case, by association, that Sauer's cultural landscape thesis fits Dilthey's experiential assumptions, in that Sauer's association of areal experience with "areal and habitat differentiation" opens that door; but the evidence is lacking as to specific borrowing.[33] Dilthey's question was not of idiographic description versus nomothetic explanation, but rather, what was it *like* to have lived through those times? or, more elegantly, that the special task of students in the cultural studies is to relate the development of an event (or events) in its (or their) experienced context (Lukermann 1965).[34] In the sense of "place" rather than "event," that was Sauer's thesis, but Hartshorne never saw it that way and Sauer never linked his ideas to Neo-Kantian thought except in a vague reference to Vaihinger on the matter of organic analogy.

Hartshorne's discussion of the idiographic/nomothetic question in *The Nature* and *Perspective*, limited as it was to the classification of the sciences, was not different from Hettner's or Sauer's in any significant way. Why the next generation of geographers pinned the label of idiographer on him is not rational, and the evidence is to the contrary; it was a polemical tactic, not an analytic conclusion.

# Conclusions

In summation, *The Nature of Geography* was and is an important and influential work. It will become part of the corpus of the history of the discipline. Its authority was enormously influential in support of the argument of geography as areal differentiation—the chorologic science. Geography defined by different premises, whether man-land relationships, distributions, cultural landscapes, historical geography, ecology, or spatial interaction, was affected primarily in their need and generation of counterarguments for defensive purposes. *The Nature of Geography* contributed very little, positively or negatively, to geography as a systematic science beyond its period 1939-59, other than disturbing its progress for a time. We have seen the immediate response, in the middle of the century, to be one of few in numbers on the part of the profession; they were literally overwhelmed. After 1960, the response was vociferous and polemical, basically to set aside *The Nature*'s tenets. *Perspective*, predictably, was just an echo of the past.

In the history of geographic thought, *The Nature of Geography* has its place—a very prominent one but no longer the very dominant place it held for thirty years.

## Notes

1. The title is a statement of a logical fallacy common to historians and biographers. The important element in the title is not the Latin phrase but the question mark. It has been contended that the publication of *The Nature of Geography* was as much a cause as a result. This study examines both contentions.

2. The quotation is from a cover letter of Hartshorne on submitting the manuscript to Derwent Whittlesey, editor of the *Annals*. It appears in a Foreword by the editor suggesting some of the stimulus for the writing and publication of the study. The statement underestimates the importance of an earlier study by Hartshorne on sources of geographic thought, stimulated by differences of perspective with Carl O. Sauer over the nature of political geography (Hartshorne 1939, ii; 1935; 1979, 66–67; Sauer 1927, 207–10).

3. The 1961 printing added some twenty-six pages to the original 1939 edition. The 1946 edition had extensive supplementary notes with bibliography, much of which (some thirty pages) was deleted in the 1961 edition (Hartshorne 1946, 1955, 1958, 1959).

4. Dryer's review is relevant in several ways to *The Nature of Geography* and prescient in another (Dryer 1926, 348–49):

> Learning that the great philosopher Immanuel Kant had published a Physical Geography, the reviewer found ... the apparent meaning that, while history deals with the succession of events in time, geography deals with occurrences coexistent in space. During the same quest we were cheered by the discovery of the same idea expressed by the Scotch philosopher, Alexander Bain, that the foundation of geography is the conception of occupied space. The difference was like that between finding an amorphous boulder and a cut diamond.

> According to Hettner, "a general earth science is impossible of realization; geography can be an independent science only as chorology." To discover the areal connection of phenomena and their order is the only task to which geography should devote its energies.

> Geography assumes the responsibility for the study of areas because there exists a common curiosity about the subject. It existed long before the name was coined. His field being thus well established in primitive human instinct, as well as in the philosophical "space" of Kant and Bain, the geographer may proceed calmly to cultivate it. Parenthetically it may however be remarked that the thought of what super-Euclidian geometry and Einstein's relativity may do to the geographer's "space" and the historian's "time" is liable to give both a bad half-hour.

5. A convenient resumé of American disciplinary activity in the first forty years of the century is in James and Martin, *The Association of American Geographers: The First Seventy-*

*five Years 1904–1979*, particularly as to the intertwined roles of Sauer, Hartshorne, Leighly and Whittlesey (James and Martin 1978, 8–9, 51–54, 71–72, 79–81 and 107; Dryer 1920; Fenneman 1919; Sauer 1925, 1927).

6. The 1937 article by Leighly had few footnotes and was largely personal musings on the aesthetic nature of geographical literature dealing with the region, the "vain dreams of a science of regions," and the sterile ground of regional description. Robert Platt responded in 1938, pointing out that regional synthesis must have recourse to "phenomena which are incommensurable and heterogeneous," that can be described and understood successfully without explanation in strictly scientific terms (Leighly 1937, 130–31; Platt 1938, 33).

7. The observation is quoted from James and Martin, but is elaborated on in Hartshorne's 1979 memoir on the genesis of *The Nature of Geography*. Both Sauer and Leighly figure prominently in Hartshorne's bibliobiographical note (James and Martin 1978, 81; Hartshorne 1979, 63, 69).

8. Leighly's 1937 *Annals* article should be paired with his review of Gottfried Pfeifer's article (in German) on theory and method in recent American geography. Pfeifer covered both the environmentalist and regionalist issues. In the latter case, Pfeifer was critical of the American geographer's emphasis on the study of regions being "means"-oriented and, therefore, being theoretically defective, as against an "ends" orientation such as regional planning and development. Leighly's 1938 article was quite traditional as to subject matter: Ritter, Froebel, Peschel, von Richthofen, Gerland, Hettner. But it appeared in the midst of Hartshorne's work on *The Nature* manuscript and, he says, added nearly one hundred pages (printed) to the final product (Leighly 1937, 1938a, b; Hartshorne 1979, 72–73).

9. The furor in America precipitated by Leighly (and Sauer) had parallels in Britain raised by Crowe's article in 1938, "On Progress in Geography." Dickinson, armed with German, French and American landscape morphology rhetoric, responded in kind and the battle was joined. Chorology, morphology and landscape became key concepts in the arena of the traditional dualisms of systematic versus regional geography, as it had been in the nineteenth century on the continent; the English-speaking world was just catching up (Crowe 1938; Dickinson 1939; Forde 1939; Stevens 1939). The trans-Atlantic connection was verified in Finch's presidential address of 1938 where Leighly, Crowe, Pfeifer, Vallaux and Bowman are cited along with Fenneman, Hettner, Granö, A. Penck, Sauer and Kroeber. The discipline was obviously disturbed (Finch 1939).

10. Sauer's presidential address of 1940 was bitter in one instance in that he labeled the period from Barrows (1922) to Hartshorne (1939) as "the Great Retreat," a retraction of geography from its natural science roots by the renunciation of physical geography. The second retreat was that of "adaptation" and "adjustment," the attempt to create a *science* rather than a *history* of human environment (Sauer 1941, 2).

11. Klimm's review provided material for Hartshorne's 1948 paper on *Mores* as well as a collation of specific criticisms for later writings. At Berkeley, after Leighly took on the major responsibility for the "Principles of Geography" course in the 1930s from Sauer, the major concentration was on the history of American geography in the nineteenth and twentieth centuries, for example, Maury and Marsh (Klimm 1947; Speth 1981, 235–37).

12. The Schaefer paper contains a footnote on Bergmann's editorial role. Schaefer and Bergmann were colleagues and close friends, both refugees from Europe in the 1930s and associated with the Vienna Circle positivists. Schaefer's references to Victor Kraft and his review of geographic methodology stems from that same philosophic background. Hettner was generally viewed with favor by positivist philosophers interested in geography because of his earlier philosophical training and associations. Bergmann's interest in composition (morphologic) laws is spelled out in detail in his 1957 book on *Philosophy of Science* in the section on Configurations and Reductions (Bergmann 1944, 1957, 131–71; Kraft 1929; Schaefer 1953).

13. Schaefer begins the second section of his paper: "The father of exceptionalism is Immanuel Kant." Exceptionalism is basically the thesis that "geography is quite different from all other sciences, methodologically unique, as it were." History as defined by Kant is also exceptionalist, according to Schaefer. It is from that base that historicism arises, and in geography, both Hettner and Hartshorne are its practitioners. Historicism is that belief that the present is the product of its past, that by merely arranging the past events in their temporal order, a meaningful pattern will appear, and likewise geography as to spatial order. For Schaefer, a regional geography assembled in this manner is merely descriptive, without scientific value—anathema (Schaefer 1953, 63–70). On Kant, see section of this paper on Historical Geography.

14. In *Perspective,* an extended bibliography introduces the post-war research of Ackerman, Bobek, Philbrick and Ullman as worthy examples of Hartshorne's acceptance of systematic geography studies, but that "first wave" was not enough to satisfy the "second wave" of the sixties (Hartshorne 1959, 191–93).

15. Schaefer (or Bergmann) gives as examples of scientific (causal process) deterministic studies: von Thünen, Palander, Hoover and Christaller (Schaefer 1953, 80, 82, 84).

16. A convenient listing of Hartshorne's research publications can be found in the bibliographies of *The Nature* and *Perspective* in the fields of economic and political geography (Hartshorne 1939, 18–19; 1959, 192). References to the "genetic" method and classification are found throughout *The Nature* (Hartshorne 1939, 418) and in *Perspective* (Hartshorne 1959, 81–107).

17. References to "material versus immaterial phenomena" and "observable features" are scattered throughout *The Nature* and in the chapter on "Significance" in *Perspective* (Hartshorne 1939; 1959, 36–47; 1979, 67).

18. The full citation is to the Abstract, Introduction 1.C which reads (Hartshorne 1939, vii): "Since it was desirable to give full consideration to ideas seriously urged by competent geographers, these discussions are necessarily lengthy and detailed. Some readers may wish to omit those sections, totalling nearly one third of the total text, that arrive at negative conclusions. These are Sections III, part A of V, VII, IX excepting part F, and parts C and E of X."

19. This article appeared in the "old" *Encyclopaedia Britannica.* A reprint is available in a recent book of James's articles edited by Meinig (James 1956, 1970, V. 10, 154–57; Meinig 1971, 2–32).

20. *Perspective*'s first draft went to Whittlesey, but was published under the editorship of Andrew H. Clark (Hartshorne 1959, ed. note).

21. In the Foreword, Hartshorne specifically takes up the question of Hettner's role in *The Nature* and whether that book was merely derivative of Hettner's basic proposition (Hartshorne 1959, 1, 10): "To eliminate any appearance of deductive reasoning from a priori theories concerning either geography or science, this book will proceed inductively, seeking to determine the actual character of geographic work as geographers have viewed that work."

22. Despite Hartshorne's awareness that the overall critical impression of *The Nature* was that it was a defense of Hettner's statement of the logical position of geography in the total field of knowledge, he reiterates the basic methodology of inquiry of *The Nature* (Hartshorne 1959, 10–11).

23. The dilemma facing Hartshorne is quite clear in this passage. He does not really see the issue as historical, but logical. He chooses explanation over understanding (Hartshorne 1959, 8).

24. Why Hartshorne chose Kroeber, an anthropologist, to define "history" is never made clear. Kroeber carefully avoids a Kantian definition, or a process definition, or a chronological sequence definition of history. His approach is *canonical,* one might even say morphological. It speaks to *period,* not time directly as in a series of events (Kroeber 1935, 545–46; Hartshorne 1939, 129, 145, 183, 283, 417–18, 446).

25. Hettner's training at Heidelberg is not emphasized in *The Nature* nor in *Perspective* (Plewe 1982, 57). More current discussion may be found, but obliquely, in discussion of Neo-Kantian influence in chorology and historicism; see section on Neo-Kantian Geography.

26. Varenius in the Methodus of the *Geographia Generalis* (1650) links chronology, astrology and geography as "knowledge of things which is gained by force of argument or the testimony of the senses," and separate from other knowledge gained by "demonstration or based on things highly probable." While emphasizing the parallelism methodologically of the three sciences noted above, Varenius, in other contexts, associates astronomy with geography and astrology, separating chronology (history) by implication from the other three. In the next generation of geographers, the first half of the 1700s, Johann Michael Franz, a cartographer-geographer, explicitly separates history and geography: "One must observe in which different ways the terrestrial space is filled with regard to is configuration . . . . All that is described by history is here [Geography] depicted in its spatial configuration" (Varenius 1650; Lukermann 1963; Jäkel 1981).

27. The relevant lecture notes in the Einleitung are in section #4 as follows (Kant 1802): "Concerning the plan for the organization, we have to put all our knowledge in its proper position. We can refer our empirical perceptions to a position either under concepts or according to time and space, where they actually can be found." Further: "The story of what occurs at different times, that is, actual history, is nothing but a continuing geography. Thus it is one of the greatest historical defects if one does not know in which place something occurred or what the circumstances were." And vice versa.

28. In addition to the criticism of Broek's study, the work of H. C. Darby and A. H. Clark were brought into question by Hartshorne and Cumberland: "changes from time to time are the concern of history; differences from place to place are the concern of geography" (Hartshorne 1939, 177–78, 188; Hartshorne 1959, 107; Cumberland 1955, 185).

29. Hartshorne considered Broek's thesis on the Santa Clara Valley as the only complete study that carried out Sauer's 1925 outline of a landscape study; but he found it wanting and suggested "The question here is whether we would not have an even clearer picture if the study had been organized entirely on the present landscape, and in the analysis of each of its parts had utilized the appropriate material that was presented in the historical chapters" (Hartshorne 1939, 179; Sauer 1941, 4).

30. It is important to emphasize that Hartshorne's interest in the nature of geography did not just stem from the 1937 and 1938 articles and meetings. Hartshorne's intense interest in political geography and his field work in Germany in 1931–32 are the real basepoint for the research that culminated in 1939 with the publication of *The Nature* and almost wholly occupied his mind in one way or another through the publication of *Perspective* in 1959 and on to the present (Sauer 1927, 1931; Hartshorne 1979, 1988).

31. Sauer was very explicit in the "Morphology of Landscape" as to *facts, scene, landscape, observation* and *construct/concept*:

—The facts of geography are place facts; their association gives rise to the concept of landscape.
—In the sense here used, landscape is not simply an actual scene viewed by an observer. The geographic landscape is a generalization derived from the observation of individual scenes.
—The content of landscape is something less than the whole of its visible constituents.
—Beginning with infinite diversity, salient and related features are selected in order to establish the character of the landscape and to place it in a system.
—The geographer is in fact continually exercising freedom of choice as to the

materials which he includes in his observations, but he is also continually drawing inferences as to their relation.

—We cannot form an idea of landscape except in terms of its time relations as well as of its space relations.

It was on this basis that Sauer insisted that the areal experience of the culture group was what was expressed through the generic ideas of landscape and chorology (Sauer 1925, 26–28, 36).

32. The contemporary efflorescence of interest stems more from an interest in Sauer than in Hartshorne, but the context has to include Sauer's reaction to *The Nature* (Speth 1981; Entrikin 1980; 1981, 11–12; 1984, 403).

33. Sauer introduces themes of experience, habitat, and values throughout the "Morphology of Landscape" but does not seek support from the Neo-Kantian German literature (Sauer 1925, 20–21, 29–30).

34. There is a widespread literature on Neo-Kantianism by advocates as well as commentators in various encyclopedias. More specific references in their relevance to Hartshorne, Schaefer and Sauer are Lukermann 1965, 195; Makkreel 1969; Entrikin 1981, 1984.

# References

**Bergmann, Gustav.** 1944. Holism, historicism and emergence. *Philosophy of Science* 11: 209–21.

———. 1957. *Philosophy of science.* Madison: University of Wisconsin Press.

**Crowe, Percy.** 1938. On progress in geography. *Scottish Geographical Magazine* 54:1–19.

**Cumberland, Kenneth B.** 1955. American geography: Review and commentary. *New Zealand Geographer* 11:183–94.

**Dickinson, Robert E.** 1939. Landscape and society. *Scottish Geographical Magazine* 55: 1–15.

**Dryer, Charles R.** 1920. Genetic geography: The development of the geographic sense and concept. *Annals of the Association of American Geographers* 10:3–16.

———. 1926. The nature of geography. *Geographical Review* 16:348–50.

**Entrikin, J. Nicholas.** 1980. Robert Park's human ecology and human geography. *Annals of the Association of American Geographers* 70:43–58.

———. 1981. Philosophical issues in the scientific study of regions. In *Geography and the urban environment*, Vol. 4, ed. D. T. Herbert and R. J. Johnston, pp. 1–27. London: John Wiley.

———. 1984. Carl O. Sauer, philosopher in spite of himself. *Geographical Review* 74: 387–408.

**Fenneman, Nevin M.** 1919. The circumference of geography. *Annals of the Association of American Geographers* 9:3–12.

**Finch, Vernor C.** 1939. Geographical science and social philosophy. *Annals of the Association of American Geographers* 29:1–28.

**Forde, C. Daryll.** 1939. Human geography, history and sociology. *Scottish Geographical Magazine* 55:217–35.

**Hartshorne, Richard.** 1935. Recent developments in political geography. *American Political Science Review* 29:785–804, 943–66.

———. 1939. The nature of geography. *Annals of the Association of American Geographers* 29:171–645. Also 1946, *The Nature of Geography*, and 1961 revised ed. Lancaster, PA: The Association of American Geographers.

———. 1948. On the mores of methodological discussion in American geography. *Annals of the Association of American Geographers* 38:113–25.

———. 1955. "Exceptionalism in Geography" re-examined. *Annals of the Association of American Geographers* 45:205–44.

————. 1958. The concept of geography as a science of space, from Kant and Humboldt to Hettner. *Annals of the Association of American Geographers* 48:97–108.

————. 1959. *Perspectives on the Nature of Geography.* Chicago: Rand McNally.

————. 1979. Notes toward a bibliobiography of the *Nature of Geography. Annals of the Association of American Geographers* 69:653–76.

————. 1988. Hettner's exceptionalism—fact or fiction. *History of Geography Journal* 6: 1–4.

**Jäkel, Reinhard.** 1981. Johann Michael Franz, 1700–1761. In *Geographers: Biobibliographical studies,* Vol. 5, ed. Thomas W. Freeman, pp. 41–48. London: Mansell.

**James, Preston E.** 1970. Geography. *Enclyclopaedia Brittanica,* Vol. 10, pp. 154–57. Chicago: Britannica.

————, **and Martin, Geoffrey J.** 1978. *The Association of American Geographers: The first seventy-five years 1904–1979.* Washington: Association of American Geographers.

**Kant, Immanuel.** 1802. *Physische Geographie.* Berlin: Berlin Academy of Sciences.

**Klimm, Lester E.** 1947. The "Nature of Geography": A commentary on the second printing. *Geographical Review* 37:486–90.

**Kraft, Viktor.** 1929. Die Geographie als Wissenschaft. *Enzyklopädie der Erdkunde,* Teil; *Methodenlehre der Geographie,* pp. 1–22. Leipzig.

**Kroeber, Alfred L.** 1935. History and science in anthropology. *American Anthropologist* 37:539–69.

**Leighly, John B.** 1937. Some comments on contemporary geographic method. *Annals of the Association of American Geographers* 17:125–41.

————. 1938a. Methodologic controversies in nineteenth-century German geography. *Annals of the Association of American Geographers* 28:238–58.

————. 1938b. Theory and method in recent American geography. *Geographical Review* 28:679.

**Lukermann, Fred.** 1963. The intimate relation of chronology, geography and astrology according to Bernhardus Varenius. *Annals of the Association of American Geographers* 52:606.

————. 1965. Geography: De factor or de jure. *Journal of the Minnesota Academy of Science* 32:189–95.

**Makkreel, Rudolph A.** 1969. Wilhelm Dilthey and the Neo-Kantians: The distinction of the *Geisteswissenschaften* and the *Kulturwissenschaften. Journal of the History of Philosophy* 7:423–40.

**Meinig, Donald W.** 1971. *On geography: Selected writings of Preston E. James,* ed. Donald W. Meinig. Geography Series No. 3, Syracuse: Syracuse University Press.

**Platt, Robert S.** 1938. Items in the regional geography of Panama; with some comments on contemporary geographic methods. *Annals of the Association of American Geographers* 28:13–36.

**Plewe, Ernst.** 1982. Alfred Hettner, 1859–1941. In *Geographers: Biobibliographical studies,* Vol. 6, ed. Thomas W. Freeman, pp. 55–64. London: Mansell.

**Sauer, Carl O.** 1925. The morphology of landscape. *University of California Publications in Geography* 2:19–53.

————. 1927. Recent developments in cultural geography. In *Recent developments in the social sciences,* ed. E. C. Hayes, pp. 154–212. Philadelphia: Lippincott.

————. 1931. Cultural geography. In *Encyclopedia of the Social Sciences* 6:621–23. New York: Macmillan.

————. 1941. Foreword to Historical Geography. *Annals of the Association of American Geographers* 31:1–24.

**Schaefer, Fred K.** 1953. Exceptionalism in geography: A methodological examination. *Annals of the Association of American Geographers* 43:57–84.

**Speth, William W.** 1981. Berkeley geography, 1923–33. In *The origins of academic geography in the United States,* ed. B. W. Blouet, pp. 221–44. Hamden, CT.: Shoestring Press.

————. 1986. Historicism: The disciplinary world view of Carl O. Sauer. In *Carl. O. Sauer—A tribute,* ed. Martin S. Kenzer, pp. 11–39. Corvallis, OR: Oregon State University Press.

Stevens, A. 1939. The natural geographical region. *Scottish Geographical Magazine* 55: 305–17.

Varenius, Bernhardus. 1650. *Geographia Generalis.* Amsterdam: Elzevier.

# *The Nature of Geography* and the Schaefer-Hartshorne Debate

GEOFFREY J. MARTIN

Department of Geography, Southern Connecticut State University,
New Haven, CT 06515

T*he Nature of Geography* was published fifty years ago, and fifty years after George G. Chisholm's *Handbook of Commercial Geography* appeared. In that same year, television made its official debut, the New York World's Fair was opened, and World War II began.

Hartshorne has explained how he came to write *The Nature* (1979). He realized during the 1930s that there was little comprehension, accord or agreement among American geographers concerning the nature of their field. At the 1937 annual AAG meeting in Ann Arbor, Michigan, Hartshorne had lunch with Derwent Whittlesey, editor of *The Annals,* and criticized a recent *Annals* paper by John Leighly on "Some Comments on Contemporary Geographic Method" (1937). Hartshorne believed that Leighly treated methodological questions seemingly without knowledge of previous methodological studies. Upon his return to Harvard, Whittlesey wrote to Hartshorne inviting him to make "a statement to that effect—it can be brief—followed with a bibliography that will make the point."[1]

During the next eighteen months, Hartshorne wrote and Whittlesey rendered commentary. The manuscript grew ever larger. By April, 1938, a manuscript of 61 pages entitled "The Nature of Geography" was sent to Whittlesey; by July it was 194 pages. Hartshorne left Boston in August by ship for Europe, where he continued to work on the manuscript. By April, 1939, it exceeded 600 pages. His knowledge of German, his stay in the library of the University of Vienna, and his association with Sölch are well known. Whittlesey was enthusiastic about the manuscript and invited J. Russell Whitaker to render a report. Whitaker offered, "Here is a great book, a landmark in the history of geographic thought . . . a truly major contribution."[2] He provided suggestions for improvement which Hartshorne accepted.

With publication came letters of appreciation from a number of geographers, and reviews in North American and European journals, and an invitation from V. C. Finch and G. T. Trewartha, on the occasion of the 1939 annual AAG meetings, to accept a position with the Department of Geography at the University of Wisconsin. Reviews in the literature provided confirmation that the work was needed and that it had been accomplished in a most rigorous and scholarly manner (Wright 1941; Myres 1940).

Letters to Hartshorne revealed a keen sense of his accomplishment. John K. Wright wrote, "I think it's a great piece of work. . . . You have done geographers a great service by presenting such an impressive survey of geographical thought." Stephen B. Jones wrote, "I am loudly applauding your Himalayan contribution." Jan O. M. Broek said, "It is undeniable that this is the first time that a major attempt has been made in the English language to give a broad treatment of geography as a science." George B. Cressey told him, "I am going to be indebted to you for many years to come. It is the kind of thing that we are greatly in need of doing and I need not remind you that you have placed the whole profession under deep obligation." Other supportive letters arrived from Nels A. Bengston, Vernor C. Finch, Robert B. Hall, John B. Leighly, T. Griffith Taylor, and Stephen S. Visher. Letters of appreciation came from overseas, notably from Alfred Hettner, Victor Kraft, Alfred Philippson, Otto Schlüter, Johann Sölch, and Leo H. Waibel in Germany and Walter Fitzgerald in England.[3]

Wellington "Duke" Jones, who had studied with Hettner at Heidelberg in 1913 and studied and co-authored an article with Sauer while pursuing doctoral study at Chicago, had later supervised Hartshorne's dissertation at Chicago. Keenly interested in the history and methodology of the field, he stood at a most advantageous confluence of circumstances. Jones read *The Nature* in its entirety, sending Hartshorne more than thirty-five handwritten letters pursuant to many of his sessions with the publication. Thrilled that at last this void in the literature had been filled, he wrote Hartshorne "You are the only American geographer, living or dead, who could have done this job."[4]

Assimilation of the book into the professional skein was perhaps slowed by the outbreak of war in Europe and the redistribution of professional geographers in the U.S. Graduate classes in which the book might have been widely read were suspended and not resumed until peace-time. The work was initially published in two ample numbers of the *Annals* in 1939 and subsequently, that same year, in book form. It was revised over the years, published in a Japanese translation (1957), and reconsidered in the companion book *Perspective on the Nature of Geography* (1959). *The Nature* was read in the United States and abroad by professional geographers and other disciplinarians and discussed in geography seminars. Previous attempts to understand the origins, development, methods, and posture of U.S. geography included articles by W. M. Davis (1903, 1904, 1932), W. S. Tower (1910), N. M. Fenneman (1919), C. R. Dryer (1920), A. E. Parkins (1934), and Isaiah Bowman (1934). It was the first time in the English-speaking world that there had been presented a book of this genre, dealing with what geographers past and present had thought geography was, had been, or should be.

In 1953, Fred K. Schaefer's "Exceptionalism in Geography: A Methodological Examination" was published in the *Annals*. This was, he claimed, the first challenge in fourteen years mounted against the views expressed in Hartshorne's *The Nature*. Challenges of different orders of magnitude, however, had been levied previously by Sauer (1941), Whittlesey (1945), and Ackerman (1945).[5]

Hartshorne published a letter in the *Annals* (1954) asserting that Schaefer's article consisted in large part of radical misrepresentations of what he had written in *The Nature* or writers he had quoted, notably Kant, Humboldt, and Hettner. He would present detailed demonstration of these charges later, but his letter, he claimed, would "serve as a *caveat* to students of methodology and the history of geographic thought" (109). Hartshorne also prepared "a presentation for submission to the *Annals*, together with whatever footnote corrections of specific errors in the recent article were considered necessary and appropriate" (1954, 109). This he did in a paper published in the *Annals* (1955) as "'Exceptionalism in Geography' Re-examined."

At that juncture, Hartshorne thought that the damage done to the science of geography by Schaefer should have been corrected. He had already planned to write a second paper that would clarify methodological questions raised by others since the publication of *The Nature*. This work was published by the AAG under the title *Perspective on the Nature of Geography* (Hartshorne 1959). In the course of preparing the *Perspective*, "The Concept of Geography as a Science of Space, from Kant and Humboldt to Hettner" emerged incidentally (Hartshorne 1958). Further correspondence and published writing on this subject have continued into the present. Hartshorne's most recent formal statement on this matter was an invited presentation, later published as "Hettner's Exceptionalism—Fact or Fiction" (Hartshorne 1988), given before a 1983 West Lakes AAG symposium on "Exceptionalism in Geography—Thirty Years after Schaefer."[6]

## Beginnings of the Debate

The purpose of this essay is to synthesize and render an orderly account of the Schaefer-Hartshorne debate. The "debate" was inaugurated when Schaefer submitted a manuscript entitled "Exceptionalism in Geography: A Methodological Examination" to Henry M. Kendall, editor of the *Annals*. From notes in Schaefer's papers, it seems that the manuscript was submitted on 4 December 1952.[7] Kendall sent it to three members of the editorial board, Stephen B. Jones, Edward Ullman, and Clyde Kohn.[8] On 14 April 1953, Kendall wrote to Schaefer, "Your manuscript has stirred up so much discussion among the members of the Editorial Board that I have found it very difficult to come to a positive decision." Kendall cited seven criticisms from Jones and four from Ullman.[9]

The criticisms from Jones, as stated by Kendall in his letter, included that Schaefer "uses many emotionally-loaded words without defending their validity"; "Examples of substantive studies illustrating his points are lacking throughout"; "There is no real discussion of . . . the time-dimension of geography"; "The paper seems to be full of assertions rather than demonstrations"; "The paper contains a mountain of criticisms of everybody from Kant on down, but gives birth only to a mouse of constructive thought."

The criticisms from Ullman (again, quoting Kendall's letter to Schaefer) included that Schaefer had written "'some fundamental ideas remained unchallenged for decades though there is ample reason to doubt their power.' This may or may not be true, but he certainly would do well to give examples"; "He writes that Hartshorne's work has not been challenged. . . . Is it our business to tell Schaefer to read certain articles prepared by Ackerman, and perhaps others?" "He writes that 'Spatial relationships are the ones that matter in geography, and no other.' I do not agree. Schaefer begins to ask for 'laws concerning spatial arrangements.' On this business of laws, he seems to me like a high school sophomore. . . . I wonder if he has discovered any laws regarding spatial arrangements? One or several examples would be helpful. The above are merely a few examples of sharp disagreements I have with the author of the enclosed paper."

Kohn's reply was not cited by Kendall, and is otherwise unavailable. In his letter to Schaefer, Kendall also wrote:

> What I should like you to do is to give some thought to exerpts of the written critiques which follow. After you have considered them, I should like to know whether you care to revise your paper. If you still feel that revision is unnecessary and you wish the paper published, I shall be happy to receive it again.

Schaefer replied to each of the criticisms in a 1500-word statement, but he did not make any change in his manuscript. He wrote:

> Let me sum up what I think are the main points of my essay:
>
> (1) I believe I have challenged successfully the assertion that geography and history are fundamentally different from all the other scientific fields, and therefore in need of a special methodology.
> (2) The concept that geography cannot be a science and that it has to remain, probably for ever, a field of naive description or taxonomy, is opposed and shown to be untenable.
> (3) The article objects to the concept of geography as an encyclopedic description that must deal with everything in the region and shows, incidentally, the historical origin of this misconception. Humboldt's concept of geography as a science was confused with his own and Kant's cosmography.
> (4) The article tries to *outline* geography as a science. As a result of that position it sees a number of basic concepts in a light that is different from that of the traditional one. A number of fundamental concepts such as dualism, description vs. explanation, science vs. Wissenschaft, determinism, etc., were criticized, partly defined or redefined.[10]

On 9 May 1953, Kendall wrote to Schaefer accepting the paper, but added that:

> In reading your answers to some of the points made by the reader, I feel that you misinterpret. . . . I feel constrained to say that I believe you have made numerous assertions without proof, or sufficient substantiation. . . . My concept of editorial duties does not lead me to change an author's way of saying what he has to say.[11]

Kendall rendered the manuscript unchanged in galley, then sent it to the author, but Schaefer died on 6 June 1953, at age 48, of a second heart attack. A colleague and friend at the University of Iowa, Gustav Bergmann, who had already been acknowledged by Schaefer as one who "has kindly read the manuscript and made many valuable suggestions," was asked to read the galley-proof (see Schaefer 1953, 226, forematter to notes). A belief held by some contemporaries is that Bergmann influenced Schaefer considerably and might even have written parts of the essay itself, either before Schaefer's death or after, when the manuscript was in the galley stage. Years later, Hartshorne sent a statement to Bergmann, entitled "Notes on the History of Geographic Thought in the United States since 1921: *The Schaefer article, 1953.*" It was accompanied by a letter in which Hartshorne wrote: "If I have mis-stated or misunderstood you in any respect in the enclosed, I would wish to be corrected."[12] In his enclosed "Notes . . . ," Hartshorne said:

> His (Schaefer's) footnote, expressing indebtedness to a colleague in the philosophy department at Iowa for having read the manuscript and made many valuable suggestions must be dismissed, since that colleague has stated that he had not read the manuscript but had only performed, later, the "trivial" service of reading the proof.

Bergmann did not reply so the matter must rest.

Publication of Schaefer's paper raised an issue of editorial prerogative in that Kendall decided to publish notwithstanding the objections of Jones and substantial criticism from Ullman. Jones wrote to Ullman:

> I have my reservations about the Schaefer paper. I cast my vote against publication in its present form (now I'm as immoral as you) and suggested that Henry tell the author to lay it aside until he had done some substantive work to carry out his theory. But now Schaefer is dead and the illustrative study presumably is not to be. . . . The sixty-four bottle-top question is whether the more idiographic subjects can move toward the nomothetic to any significant degree. That's what I wanted Schaefer to demonstrate by substantive work. It's not enough to say they *ought* to.[13]

Hartshorne had not been made aware that the Schaefer article was to be published in the *Annals.* He read it in late October, 1953, and wrote to Kendall enclosing "statements of the major errors and mis-statements concerning the view of the writers discussed in that paper," adding "It . . . requires a corrective statement in the *Annals.* . . . I think this is an editorial question."[14] A few days later, he wrote again to Kendall that Schaefer's article was:

> a palpable fraud, consisting of falsehoods, distortions, and obvious omissions. It is not a question of disagreements . . . but of misrepresentation. . . . *The Annals* is the presentation of American geography to our colleagues in other countries. German geographers who know the history of their field will wonder that such nonsense could be published about Hettner's writings and perhaps most of all be amazed at the manner in which Humboldt is misrepresented to destroy what he clearly stated to be the fruition of his scientific work.[15]

Kendall, a faculty member of Miami University, Oxford, Ohio, was at the time "going over *The Nature of Geography* in my seminar." He offered to give Hartshorne space in a new section which was to appear for the first time in the

December, 1953, issue "directly designed for comment on articles appearing in the *Annals.* I shall allow you as much space as I possibly can. . . ." Kendall referred to "the difficulty of ascertaining the exact meaning of the premises, due to the barriers of language" . . . and wrote that there were "several places, for example, in *The Nature of Geography* where my own understanding of the original meaning, particularly regarding statements in French, would differ widely from yours."[16]

This drew another letter from Hartshorne,[17] requesting references to those "places." Kendall claimed that he had "more work stacked up than I can reasonably handle."[18] Additional correspondence to Kendall revealed Hartshorne's tenacity and desire for accuracy in the matter; it also produced a suggestion that any time a paper was confronted with a serious "challenge," the author would be allowed to read the challenging paper, and then write to the editor or publish an answer in the same number of the periodical.[19] Further correspondence by Hartshorne concerning the matter of editorial policy was sent to Jones and to Walter Kollmorgen (who had succeeded Kendall as editor of the *Annals*).[20]

## Anatomy of the Debate

To understand the "debate," it is certainly desirable (if not necessary) to have some knowledge of that now receding genre in American geography, the history of geographic thought (Kenzer 1988). Schaefer proposed in his article that some of the views Hartshorne expressed in *The Nature* (views derived largely from the writings of Hettner) were preventing geography from developing as a science. Schaefer argued that Hettner had borrowed from Kant the fallacious notion that geography was like history and so methodologically different in a number of specific respects from all other sciences. For this position, he coined the term exceptionalism. The all-important thesis in Schaefer's paper was the issue whether geography studied individual areas or sought generic principles or scientific laws. While Hartshorne had introduced into American geography the issue of the idiographic and nomothetic approach, first noted by Wilhelm Windelband (1894), it was the Schaefer-Hartshorne debate which brought the issue to the fore.[21]

Hartshorne, following Hettner, had written that the establishment of generic concepts and laws (nomothetic) was not the end purpose of science, but rather "the means to the ultimate purpose—the knowledge of actual reality" (1939, 379; see discussion, 379–84) of untold numbers of individual phenomena. But insofar as any phenomenon differs from all others of the same category, for example, any person or place, unique characteristics can only be considered as a distinctive case, that is, idiographically. Both methods of study are important in varying degrees in all sciences, except mathematics, theoretical physics, and chemistry.

Schaefer first asserted that Hettner's work was solely idiographic, did not seek to establish scientific laws, and adopted a sterile historicist posture. Carl O. Sauer was cited as the outstanding representative of historicism in America. Hartshorne's position was also characterized as idiographic. Schaefer, in a reversal of his previous assertion, then cites cases in which Hettner supports the nomothetic position. Of this Hartshorne has written (1988, 2):

> This leads to a double-barrelled conclusion knocking out both targets: Hettner is found to have been confused and vacillating, advocating at different times and different places the idiographic or the nomothetic conceptions; and I am found responsible for misleading American students by using quotations from Hettner that give only the idiographic side of his position.

In "Exceptionalism in Geography Re-examined," Hartshorne reexamined Schaefer's paper line by line. The criticism is well known. It was an éclaircissement and, as Hartshorne made clear, in no way personal (1955). He set about revising Schaefer's essay and found it necessary to make many dozens of corrections. Hartshorne revealed the lack of precise citation, the exploitation of the paraphrase devoid of reference, and the summary of extended passages controverting the meaning of the author. Hartshorne found the essay obreptitious. He wrote (1955, 242–43):

> The pattern of these falsifications and fabrications is not one of chance, resulting from undirected carelessness; rather they cause writings used in support of the views advocated in the critique to fit those views while writings under attack are distorted to express views the reader can be expected to oppose.

Neither Hettner nor Hartshorne had derived their work from the philosophy of Kant.[22] In two letters Hartshorne received from Hettner, the latter declared and otherwise indicated that Hartshorne understood him.[23] Heinrich Schmitthenner (1941, 461) wrote that *The Nature* was "in step in well nigh all parts with Hettner's methodology, indeed is grounded entirely thereon." Hartshorne could provide numerous citations in both his and Hettner's work for embracing both the idiographic and nomothetic positions. "With respect to geography the inclusion of both approaches is stated in no less than fifteen places in my first book and many times since." (Hartshorne 1939, 382–86, 391, 396, 431, 434, 446, 458, 464, 466–68, 1955, 232, footnotes 76, 77, 1959, 157–65). This is the point that Myres made in reviewing *The Nature* (1940, 398–99):

> is a careful and sensible account of the twofold aims and procedure of geography, as a science "systematic" and "descriptive" at the same time and inevitably. The balance of interest in one aspect or the other has varied with the personalities of its exponents, and the progressive maturity of the study; but the scientific objective is the same "understanding of reality," and the true use of "systematic" universals is in drafting our descriptions of the unique.

Although Hartshorne was later criticized by those who felt that he should not have been so precisely severe in his criticism, it is significant that no one opposed the detail and substance of that criticism. He sent the "re-examination" article to Whittlesey, Kendall, S. B. Jones, and others on the *Annals* board.

Whittlesey wrote Broek, chairman of the AAG Publications Committee, urging that Hartshorne's article:

> be published in the *Annals* as the number one article in whatever issue it appears. Hartshorne proves beyond question that the article should never have been published. . . . It is too bad that any paper appearing in the *Annals* should be so far from scholarly that it can be taken apart as Hartshorne has done.[24]

In the same letter, Whittlesey urged submission of a manuscript "to critics expert in the fields of manuscripts submitted" and found this system (established by R. E. Dodge in 1911) much more satisfactory than submission to a publications committee. Following his retirement as editor of the *Annals*, Kendall wrote his successor, Kollmorgen, that while he felt the Schaefer paper was "thin in places," he also felt that it would be useful and would perhaps enliven "what to many had become a rather dull publication." Kendall then suggested "that Dick write a critical examination to be allowed as much space as Schaefer was allowed."[25] Kollmorgen discussed the proposal with his editorial board and corresponded with Hartshorne. Hartshorne wrote him back that:

> In whatever sense it is possible for a learned journal to commit a crime . . . The *Annals* has committed a crime unparalleled in its history. . . . The crime is not expressed by Ed's [Ullman] phrase "Schaefer versus Hartshorne," as in a legitimate debate, but as Schaefer falsifying Kant, Humboldt, Hettner and Hartshorne, not to mention Platt and others.

Hartshorne suggested that the method of evaluating papers submitted to the *Annals* for publication be reconsidered, then reaffirmed that he wished to write two separate essays, preferably to be published in different numbers of the *Annals*. Hartshorne stated that his first paper, "'Exceptionalism in Geography' Re-examined," would "reduce the Schaefer business to nearly zero," and that his second paper would answer questions which went beyond *The Nature*:

> Hence I conclude it proper to discuss these questions as though the S. [Schaefer] paper had never been written, endeavoring to supply answers to Steve, Ed, Van Cleef (two recent papers), Baulig, Kimble, and others. In part this paper is re-arrangement and clarification of things separated in *The Nature of Geography*; in part it represents definite changes in what is written there . . .

Hartshorne's second "paper" was published as *Perspective on the Nature of Geography*. He also wrote "The Concept of Geography as a Science of Space, from Kant and Humboldt to Hettner" (1958). The "Re-examination" paper constitutes an essential part of "the Schaefer-Hartshorne debate." An essay unique in the history of the *Annals*, it provoked a strong reaction.

# Theoretical Geography and the Science of Space

William W. Bunge, who had begun doctoral study at the University of Wisconsin, moved to the University of Washington and became part of an innovative group of young scholars that in 1955 included Brian Berry, John Kolars, Richard Morrill, John Nystuen, and Waldo Tobler and (in 1956) Michael Dacey, Arthur Getis, Robert Mayfield, and Ronald Boyce. It was a remarkable assemblage of young persons who were led and inspired by William L. Garrison and were endowed with energy, intelligence and belief in the scientific legitimacy of a mathematically-minded methodology. The University of Washington provided an environment in which meaningful intellectual growth of a new genre could, and did, take place. Donald Hudson and Edward Ullman were the sentinels and Garrison was the leader. Richard Morrill (1983, 59) has written: "It was easy to personify Richard Hartshorne, whose work we studied in detail, as what we struggled against. We found heroes—notably Schaefer and Christaller—and many villains." In this setting, Bunge wrote his doctoral dissertation on theoretical geography under the guidance of Garrison, who was breaking a new trail and who led a group sometimes referred to as "the space cadets." Garrison was enthusiastic about the work of Schaefer. In a note to Ullman, he wrote:

> I was and still am excited by Schaefer. Now you may present me with formal proofs (1) that all German geographers are deaf, dumb, and unable to write and (2) that Schaefer was cruel to little children, and I would still be excited by Schaefer. Excited simply because Schaefer seemed to know in some crude way of the world of science of which geography is a part.[27]

Bunge later recalled (Dow 1976):

> Garrison was able to get us the Ph.D. and no one else could. So everybody gathered there and it was a ragtag bunch. . . . When I got my degree at Washington, Bill Garrison said "Well, this just proves that Wisconsin has higher standards," and I think there is something to that. . . . But he enabled us to get our Ph.D. in mathematical geography, and that is not nearly acknowledged enough. . . .

After completion of the doctorate, Bunge accepted a position at the State University of Iowa where he taught most of the content of his dissertation, published in 1962 as *Theoretical Geography* (Bunge 1962). "This discipline led to drastic changes in almost every chapter" (1962, viii). Bunge notes in his Acknowledgements (1962, viii):

> Each chapter owes individual debts. The chapter on methodology was written with extensive aid. Professor Gustav Bergmann of the Department of Philosophy of the State University of Iowa was especially valuable since he was a close friend and intellectual associate of Professor Fred Schaefer in Schaefer's classic work, "Exceptionalism in Geography: A Methodological Examination." Such help assured the representation of Schaefer's position with unusual accuracy.

Bunge's work, dedicated to Walter Christaller, constituted a quest for theory in geography. It was consonant with the spirit of investigation in the Department of Geography at the University of Washington in the late 1950s. Ullman wrote of the book, "although naive in some places, [it] is a milestone in geographical thinking and has been reviewed favorably all over the world by leading scholars."[28]

It was written at a time when there was an emerging national and international interest in the history and philosophy of science. Ackerman (1963, 430) reminded the geographic community of the role of science in the fifty years since 1910: "This panorama of glorious scientific achievement." Geographers were beginning to be aware of the advancing front of science, and wanted to share its benefactions. Within geography the work of Christaller was being revisited, and that of Ackerman, Crowe, Garrison, Hägerstrand, Hoover, Isard, Kendall, Lösch, and Warntz was beginning to excite the attention of especially those who wished to redirect American geography. These young scholars sought a "new geography," which was not in opposition to what had been published in *The Nature*, perhaps without realizing that there had been previously several "new geographies" in the history of American geography. The "new geography" represented growth and was a part of the process of disciplinary evolution. Those in quest of a departure from what was considered to be a regional "mainstream" began to read the works of Habermas, Kuhn, Lakatos, Popper, Stewart, Toulmin, Zipf, and others. The quest for a theoretical geography larded with conceptual nuggets was facilitated by mathematical and model constructs and studied by an energetic, intelligent, and committed group of young scholars who would leave their imprint on the discipline.

In retrospect, it does seem unfortunate that this group of "new geographers" did not study in depth the history of their own field. Had they done so, they would have realized that the quest for theory, the use of mathematics, and model building was not new to the field. Hugh Robert Mill had scribed his model of geography in relation to other subjects in 1892.[29] Davis had constructed a more sophisticated multi-dimensional model, "A Scheme of Geography," in 1903. These two examples were perhaps those best known to American geographers in the first half of the twentieth century. Davis had written of theory and "theorizing as an essential part of investigation in geography, just as in other sciences" (1900, 157–70). Elsewhere Davis revealed awareness of the distinction between "the quantitative and qualitative" and:

> the tyranny of theory; truly there is a great danger; I have experienced it. But I do not think the danger is at all peculiar to . . . geographical study. It occurs in all sorts of scientific study.[30]

Geographers had been quantifying much earlier than is commonly supposed; examples are provided by Grove K. Gilbert (1914), John K. Rose (1936), and, in England, Maurice G. Kendall (1939). The late 1940s also witnessed the beginnings of experimental research in the genre of geography as social physics.[31,32]

It is largely "A Geographic Methodology," the first chapter of Bunge's *Theoretical Geography*, that is relevant to the "debate." Bunge criticized Hartshorne's comprehension of unique, accusing him of confusing "*unique* with *individual* case" (1962, 9) and of possessing a "crippling" attitude (10). Schaefer, Bunge pointed out, "has a strong grip on the problem of uniqueness," and:

> Symptomatically, throughout Schaefer's work runs the generic term spatial while Hartshorne uses the idiographic word place. The space versus place dispute is a

direct consequence of their positions on general versus unique. Hartshorne is pes-
simistic as to our ability to produce geographic laws, especially regarding human
behavior. Schaefer has done us a great service in sweeping away our excuses and
thereby freeing us from self-defeat (1962, 12).

Hartshorne sought to discover what geographers considered geography to
be, and for this purpose reached back in time, while Schaefer was concerned
with what geography could be. Bunge later changed his mind on the issue of
uniqueness (1979, 173):

> Hartshorne is correct about the uniqueness of locations. Considering that I have
> published under the title "Locations Are Not Unique" and that *Theoretical Geography*
> is an unmitigated attack on uniqueness, the necessity publicly to admit to my printed
> error was not without pain.

There was another area of disagreement between Bunge and Hartshorne: the
role of history in scholarship. Bunge wrote to Hartshorne (ca. 1 June 1959): "I
do not care about the historical scholarship. I consider it irrelevant. History, as
conducted in geographic methodological discussions in general, can prove anything and therefore proves nothing."[33] While this statement is curious, given
that most of Schaefer's paper consisted of historical scholarship, it reveals a
distinctive point of view.

In Bunge's *Appendix to Theoretical Geography* (1966), he sought to reveal something of the meaning of "spatial relations." This was the term Schaefer had
used without defining in his oft-quoted "Spatial relations are the ones that
matter in geography, and no others" (1953, 228). Ullman's criticism of the
Schaefer manuscript includes disagreement with that statement. He wrote, "If
I point out that certain people build with stone rather than wood because the
former material is at hand and the latter not, am I making an observation that
has no place in geography?"[34] When Schaefer received the criticism from Kendall, he replied:

> (this) proves my point! "If certain people build with stone rather than wood because
> the former material is at hand and the latter not . . ." is a clear statement of spatial
> relations, i.e. of the spatial coincidence of the occurrence of a certain type of ar-
> chitecture with the occurrence of lumber or stone respectively.[35]

If this statement is correct, then geographers J. F. Chamberlain (Los Angeles
Normal School) and Mark S. W. Jefferson (Michigan State Normal College,
Ypsilanti) were among the leading practitioners of spatial relations in the earliest
years of this century. In the absence of a measured statement by Schaefer on
this matter, Bunge addressed the subject by way of five subheadings: "The Call
for a Science of Geography," "The Subject Matter of the Science of Geography,"
"The Application of Geometry to Geography," "The Breadth of Theoretical
Geography," and "Spatial Prediction." The whole forms an interesting and
energetic excursus and prelude to the topic "spatial relations."

Bunge also wrote "Fred K. Schaefer and the Science of Geography," published
in the Harvard Papers in Theoretical Geography Series (1968). The Preface was
written by William Warntz, then Director of the Laboratory for Computer Graph-

ics and Spatial Analysis and Professor of Theoretical Geography and Regional Planning at Harvard University. Noting that none of the journals had carried an obituary notice of, or tribute to, Schaefer, Warntz wrote:

> Schaefer's work, challenging as it did the vested intellectual interests in geography and the academic empire built upon them, marked a turning point in the science of geography. Modern Theoretical Geography, as an emerging discipline in American universities, clearly must recognize that its strongest impetus came from Schaefer (1968, i).

Warntz added that Bunge had "compiled a long list of rejection notices over the years in his attempts to have published a tribute to Schaefer" and asserted that "Schaefer, especially as championed by Bunge, is still considered as a threat by those still advantageously-positioned survivors from geography's earlier establishment" (1968, i). Bunge's essay, "which was widely circulated among geographers in the 1960s," was the "most prominent of several previous mimeographed versions" and was reprinted "in a somewhat condensed form" in the seventy-fifth anniversary issue of the *Annals* (Hudson 1979, 128). Each of these versions provided biographical data concerning Schaefer that is nowhere else available.

## Schaefer and the University of Iowa

Schaefer was born in Berlin on 7 July 1904, attended public school from 1911–18, then became an apprentice metal worker (1918–21) and Secretary of the Trade Union Youth Section of the Social Democrats (1921–25). In 1925 he entered Kaiser Friedrich Realgymnasium, graduating with distinction in 1927. Schaefer returned to studies at night school and then attended the Deutsche Hochschule für Politik where he studied political science and political geography. From 1928 to 1932, Schaefer attended the University of Berlin, continuing into postgraduate studies. With the coming to power of the Nazis, Schaefer, an active Social Democrat, was apparently imprisoned in a concentration camp.[36] Harold McCarty relates (Dow 1971b, 5) "He got out on a skiing permit to Switzerland, from where he went to France, and from France to England. In England he got in touch with the Quaker Resettlement Group." In 1938 he journeyed to New York and in the following spring, to Iowa to help establish the Scattergood Rehabilitation Center, a refugee camp run by the American Friends Service Committee not far from the State University of Iowa. He apparently gave talks of a political nature throughout the state. During the war, he taught in the training program for the armed forces personnel located on the campus (McCarty 1979). When a Department of Geography was formed at the University of Iowa in 1946, Schaefer was given a faculty position. He taught a seminar in the history of geographical thought, and other courses including political geography, Europe, and the Soviet Union. This latter interest reveals itself in the only paper he read to the AAG membership, "Geographical Aspects of Planning in the U.S.S.R." (1947).

The young department was led by McCarty, who had, since 1923, been a faculty member in the commerce and non-departmental geography program, and whose work was well respected on the campus and throughout the country (King 1988).[37] Members other than Schaefer included Robert G. Bowman, Lyle E. Gibson, and George W. Hartman. John C. Hook and Duane S. Knos were soon added. A problem-solving and analytic approach was favored. McCarty encouraged the search for a body of theory which would facilitate development of the evolving science of geography. McCarty did not mention Schaefer or Schaefer's article in his "Cornbelt Connection" article (1979); neither does he mention Bergmann. Gustav Bergmann, as noted above a member of the philosophy department, was also a refugee from Nazi Germany who had been a member of the Vienna Circle, a group of logical positivists who came together in the 1920s. Bergmann and Schaefer became good friends, the latter learning from Bergmann much about positivism. Graduate students were:

> urged to take Bergmann's 'Philosophy of Science' course and 'Intro to Logic' courses. At the same time . . . to take math courses through calculus . . . statistical methods . . . and a professional seminar every Friday afternoon of every semester.[38]

Bergmann, author of *The Metaphysics of Logical Positivism*, may well have been writing a manuscript concerning the philosophy of science while Schaefer was writing his *Annals* manuscript.[39] Just how much Bergmann influenced or otherwise aided Schaefer cannot be known, but his friendship seems to have been significant.[40]

Schaefer was interested in the writings of Christaller, Lösch, and von Thünen. He was also interested in mathematics and statistics and was preparing a book on political geography. The second chapter of this incomplete manuscript was entitled "The Nature of Geography." John Hook has suggested[41] that following much exchange with Bergmann, Schaefer:

> became increasingly convinced that methodology was a major problem and decided to deal with it in a separate article rather than as a chapter in a book. One afternoon . . . he had a group of us students out to his house at which time he read to us a preliminary draft of the article. As it turned out, this preliminary draft was the way the article was actually published. I remember clearly that when he asked for our comments, we criticized the way in which he handled the historical development of geography in the first part of the article (the same part that Hartshorne criticized later). We commented that he would never let us get away with such sweeping general statements in any of our seminar papers.

> Schaefer replied that what we said was true and pointed out that this was only a preliminary draft and that he was going to patch it up later. He also pointed out that the historical development of the field was one thing, its logical structure quite something else; and he was primarily concerned with the latter.

> I do not know the date on which he finally submitted the article for publication, but it was always my impression that he had the feeling that he had to hurry up and get something in print.

As Leslie King (1979) has written, little attention was given to the essay within the Iowa department. According to his contemporaries, Schaefer tended not to develop friendly relations within the department.[42] McCarty did not view Schaefer's scholarship with much appreciation:

> Schaefer has already received more attention than he deserves. . . . His background in geography was limited (his diploma from Berlin was in economics) and he made no contribution to the literature. But he was an opportunist and an aggressive debater, rather good at finding minor flaws in other people's writing. On campus, as a refugee from the Nazis, he acquired something of a reputation as an expert on Germany during World War II, largely, he once told me, by relaying what he read in the *New York Times*. But with his German background and accent, he made it sound pretty authoritative. He was fascinated with Hartshorne's *Nature of Geography* and used it in a seminar. In retrospect, it seems quite natural that he combed it for errors in detail. When he thought he had located one, he quite naturally wrote it up. Unfortunately, however, he was becoming increasingly paranoid-schizophrenic at that time and what started as a reasoned critique wound up as a vicious personal attack on the author. This was not an isolated instance, but it was the only one that got in print. (His attack on me made the one on Hartshorne seem rather gentle!) But it ran true to type: always there was the friendly, well-reasoned appraisal followed by the searing personal attack. . . .
>
> In summary, I adjudge Schaefer's place in American geography to be of extremely minor importance.[43]

Gerard Rushton has observed that he could not recall McCarty ever making "a single reference to Schaefer," though some have assumed that McCarty's view was Schaefer's view. Rushton felt that there were two reasons for this:

> that he thought that the distribution theme in Schaefer is the same as that of de Geer's 1923 view; and that he was not comfortable with Schaefer's view that it was the pattern itself, and the relationships between patterns that comprised the explanation. McCarty was convinced that the distribution of phenomena should be the object of enquiry, but was too interested in the process of decision-making behind the patterns to be content with Schaefer's narrow criteria for explanation.[44]

Bunge's essay, "Fred K. Schaefer and the Science of Geography," contained numerous assertions which injected a personal element into the debate. Bunge described Schaefer as "a whole man, a conscious member of the human race, a scientist and an intellectual" (1968, 128), a "good prophet" (1979, 131) and proclaimed his essay a "classic work" (1962, viii). He further asserted:

> Hartshorne had long intimate conversations with Schaefer, his potential critic. Hartshorne had been a high-ranking officer in the Office of Strategic Services (spies) in the World War II. Thus, when they broke, Schaefer thought he had been personally betrayed (1968, 10–11).

The facts are at variance. Schaefer and Hartshorne met twice, once in April, 1946, at a meeting of the Midwest Economics Association, when their meeting was very brief and unrelated to methodology. The second time was in May, 1950, when Hartshorne was invited by Harold McCarty to visit the department at Iowa and to participate in Schaefer's seminar on geographic thought. Schaefer had adopted Hartshorne's *The Nature* as text for the seminar, and the session was given to students' questions concerning the book. The occasion was appreciated, for Schaefer wrote to Hartshorne "let me express my and the students' heartfelt gratitude for that session" and, of the evening social gathering, "Miller, Bergmann, Moehlman and Horn told me what a treat the evening session was.

Again, thank you."[45] The point is that Bunge's essay contains numerous inaccuracies and insinuations which, once published, diffused, seemingly to enlarge the worth of Schaefer's contribution and, in proportion, to diminish that of Hartshorne. His comments were also defamatory of Hartshorne's character. Hartshorne received a copy of the Bunge essay in March, 1969. He wrote Warntz:

> The paper is concerned in no small part, and your preface in major degree, with the relations of other American geographers to Schaefer. As the only one named in that connection, and named repeatedly, I have some interest in knowing the meaning of what is being broadcast about me.[46]

In his letter, Hartshorne requested clarification on four points. These included two items in Warntz's preface which referred to "merciless ... pressures that originate ... especially from within his own discipline" (Bunge 1968, 1) and "determined and well-organized resistance from within the United States" against publication of Bunge's book in Sweden (Bunge 1968, 2). The other two items related to what Bunge had written:

> 3. What is your view of the accusation (page 10–11) that I had "betrayed" Schaefer, with the implication that I had reported his views on Marx to the FBI and had done so because I recognized in him a potential critic of my views of geography?
> 4. Similarly, what is your view of the final conclusion of the paper which joins my alleged fury with Hitler's terrorism and McCarthyism as responsible for Schaefer's untimely death?

Warntz did not reply to this letter nor to a copy which Hartshorne sent later. Hartshorne has stated (Dow 1986) that he had been told earlier by Warntz that his major professor, an enthusiast of *The Nature*, had told Warntz, "The search for laws isn't the business of geography. No. You can't do that in geography," apparently citing Hartshorne's book. Hartshorne had not anticipated any such misuse of the book.[47]

In his essay, Bunge (1968, 11) wrote "The author of this biography freely admits that he stands in great personal hostility relative to Richard Hartshorne, as is broadly known within American geographic circles." Later, in a letter to Andrew Clark,[48] Bunge referred to Hartshorne as a "cruel man" and blamed him for his failure to pass his Ph.D. preliminary examinations at the University of Wisconsin. Hartshorne replied to a number of accusations in the Bunge letter (including the false charge that he had been a member of the CIA) explaining Bunge's failure of the "prelims":[49]

> we each had graded a question independently, then found that several, three or more, had found your answer to his question unsatisfactory, compared notes and discovered that rather than answer the question, apparently out of lack of interest in studying the field, you had simply attacked the question. There was no harm in that, but the failure to answer the question meant a failure to pass.

The point is that Bunge allowed emotion and unsubstantiated assertions to enter his writings concerning the work of Schaefer and Hartshorne. This emotional dimension enlarged, distended, and promoted the debate and its diffusion. There were a considerable number of publications in America which misrepresented the matter, most frequently asserting that Hettner and Hartshorne founded their work in Kant, and that they both proclaimed the method of geography to be idiographic. These statements borrowed relentlessly from one another with authors not returning to Hartshorne's "Re-examination."

## Reactions to the Debate

In France, Paul Claval, then of Besançon, wrote of the Schaefer-Hartshorne debate in *Essai sur l'evolution de la Geographie Humaine* (1964) and again in "Qu'est-ce que la Géographie?" which was published (in French) in *The Geographical Journal* (1967). Hartshorne (1969, 323–24) replied in the pages of *The Geographical Journal*: "I am . . . surprised to find in this article views ascribed to me (without specific references) which I certainly do not hold and I do not believe are implied in anything I have written" (Hartshorne 1969, 323). Hartshorne pointed out that his "Re-examination" paper did not criticize Schaefer's analysis, but "was concerned solely to find out and correct the misrepresentations, whether of commission or of omission, of what others had written. With these removed, I wrote, there remained *no* analyses to be discussed" (Hartshorne 1969, 324). Hartshorne sent a 2200-word statement to Claval analyzing thirteen points related to his two publications.

In British geography, the debate was also joined. R. J. Johnston stated (1987, 49), "Hartshorne argued that law-seeking is not a part of geography. According to Schaefer, however, Hartshorne disregarded one aspect of Hettner's writing which was nomothetic in its orientation, and in doing this he to some extent misled American geographers." Gregory, unknowing of circumstances, wrote (1978, 31) "Bunge goes so far as to suggest that Schaefer . . . was the victim of a calculated campaign of persecution which, if not directly by Hartshorne, certainly included him, as well as the FBI, the Office of Strategic Services (which recruited Hartshorne), and the Gestapo." Statements such as these complicated and extended the debate. Norman Graves later wrote:

> Hartshorne's firm statement that the idiographic emphasis in geographical studies must of necessity dominate over the nomothetic emphasis is based on the view that no worthwhile general laws could emanate from the kind of work that geographers were doing . . . . (1981, 86).

Hartshorne wrote to Graves, inquiring, "Where, may I ask, is this 'firm statement' to be found in my writings?" and sending references to relevant sections of his work.[50] Responding in a letter to the editor of *Terra*, (1984) Graves confirmed that his original statement was inaccurate and that Hartshorne did view nomothetic and idiographic methods as equally necessary.

## Schaefer's Legacy

In the midst of the debate, there were geographers who wanted to believe in Schaefer. A Schaefer memorial lecture was instituted at the University of Iowa in 1972. His paper was reproduced by the Bobbs-Merrill Company in their series of reprints, was reprinted in the Iowa series, "Future Knowledge," (Mullins 1973, 571), and again in *Analytical Human Geography* by Peter Ambrose who wrote:

This paper has been included despite its length and difficulty because although it was published in 1953, before the great upsurge of interest in more scientific approaches in geography, it forms a firm philosophical foundation for these approaches (1969, 24).

Hartshorne's "Re-examination" is mentioned only in passing, a pattern exhibited in a number of other books on geographic thought. This display of support reveals emotion rather than scholarship, emotion which clings to Schaefer's concept of geography as "spatial relations," even though he did not define the term or state its meaning. Hartshorne wrote of the matter: "I have not been able to discover in what way this differs from views about geography expressed by any or all of the writers he criticizes. He gives no definition of the term or statement of its particular meaning for him."[51] In an unpublished essay, Hartshorne began the consideration of Schaefer's essay as an example of an emerging genre of scientism (Hartshorne 1976).

# Summary

The debate, frequently devoid of logic, has lingered. What is often missing is recognition that Hartshorne and Schaefer had undertaken to write from differing viewpoints, with differing intent, and with different standards of scholarship. Hartshorne had attempted to bring together the views of a large number of geographers, over a period of time, and to trace evolution and definition of a viewpoint. That was why he titled his 1939 book, "The Nature of Geography: A Critical Survey of Current Thought in the Light of the Past." The elucidation of such a variety of contending viewpoints expressed in different languages at different times constituted an extraordinarily difficult task, the product of which remains the only attempt at such in the English-speaking world to this day. Of the book David Stoddart has noticed "The standards and . . . the quiet good manners of true scholarship were shown to perfection . . . in 1939" (Dow 1989, 3).

Schaefer had a different purpose. He urged geographers to search for regularity in the spatial order of phenomena, that is, the search for morphological laws. Secondly, in comprehending the impacts of spatial order, geographers would eventually work with other sciences to establish process laws. Schaefer wished to see geography develop the laws of spatial arrangements. It was unfortunate that he found it necessary to seek justification for his viewpoint in the history of geography. Here he betrayed scholarship. Henry Aay (1978, 330–33) writes:

> perhaps the most celebrated example of a paper which raised fundamental philosophical questions but supported them with erroneous, inaccurate and misleading statements from leading geographers of the past is Schaefer's "Exceptionalism in Geography." While this is an extreme example, it shows vividly how historical material may be presented in a misleading fashion to sustain philosophical arguments.

Mumtaz Khan arrives at a similar conclusion in "Schaeferomania: A Fallacy in Geographic Academia," in which he expressed concern about "the popularity of Schaefer's paper among (the) younger generation without their ever knowing about its unreliability" (Khan 1984, i). Rushton writes that Schaefer's influence in the U.S. and later in Europe was (1984, 1–6):

> unrelated to the issue about which Hartshorne and others took Schaefer to task: his interpretation of the philosophical views of Hettner. It was not the case that they accepted Schaefer's characterization of Hettner's views. . . .

The anatomy of the debate is a complex matter involving the unseen play of motif, psyche, temperament, ambition, and much else. It demonstrated the powerful role of transmitted error in the processes of shared intellection. It generated passion, logic, and illogic. It did not generate a conclusion. Geographers took from the debate according to their own disposition. There were those who considered Schaefer's work an exercise in academic legerdemain and others who sifted and sorted and felt that they derived direction from it. There remain those who continue, albeit unwittingly, to perpetuate Schaefer's inaccurate representation of the work of Hartshorne and others. The debate has grown old. It is hoped that it has assumed the dimensions of an episode in the fifty-year life-path of *The Nature of Geography*.

## Notes

### Archival Sources

(RH)—Richard Hartshorne papers (privately held). (GJM)—Geoffrey J. Martin papers (privately held). (FKS)—Fred K. Schaefer papers, American Geographical Society Archives. (ELU)—Edward L. Ullman papers, University of Washington archives. (DSW)—Derwent S. Whittlesey papers, Harvard University archives.

1. Derwent Whittlesey to Richard Hartshorne, 16 February 1938 (RH).
2. J. R. Whitaker to Hartshorne, 11 December 1939, acknowledging that he wrote the unsigned three-page report (RH).
3. John K. Wright to Hartshorne, 17 December 1939; Stephen B. Jones to Hartshorne, 15 November 1937, 25 January 1940; Jan O. M. Broek to Hartshorne, 2 December 1939; George B. Cressey to Hartshorne, 9 November 1939; Vernor C. Finch to Hartshorne, 24 February 1940; Robert B. Hall to Hartshorne, 6 November 1939; John Leighly to Hartshorne, 7 November 1939; Alfred Hettner to Hartshorne, 12 May 1940; Victor Kraft to Hartshorne, 5 July 1940; Alfred Philippson to Hartshorne, 13 October 1940; Otto Schlüter to Hartshorne, 21 April 1941; Johann Sölch to R. Hartshorne, 7 May 1940; Leo H. Waibel to Hartshorne, 15 March 1940; Walter Fitzgerald, 10 December 1945 (all RH). Other letters are misplaced, but excerpts were typed in a statement available in (RH).
4. W. D. Jones to Hartshorne, 7 November 1939 (RH).
5. Perhaps an indirect challenge also came from E. Van Cleef (1952). But this essay found argument with the practice and direction of American geography rather than with Hartshorne's book. See also Van Cleef (1955).
6. Since then the Royal Geographical Society awarded him its Victoria Medal (1984) and the AAG celebrated the fiftieth anniversary of the publication of *The Nature* at the annual meeting, Baltimore, 1989. Hartshorne had also been awarded the Charles P. Daly Medal of the American Geographical Society, 1959; the AAG Outstanding Achievement Award, 1960; and an honorary LLD by Clark University, 1971, all of which embodied to some extent recognition of the worth of *The Nature*.
7. The Schaefer papers were donated by his widow to the AGS and deposited there by W. W. Bunge.

8. Jones to Ullman, 7 October 1953 (ELU); Dow 1971a, and two telephone conversations between Martin and Kohn, 8 June and 13 July 1989. Other members of the editorial board were Ruth E. Baugh, Samuel N. Dicken, Stanley D. Dodge, Walter M. Kollmorgen, and Glenn T. Trewartha.

9. Henry M. Kendall to Schaefer, 14 April 1953 (FKS).

10. Schaefer's undated reply is available, but there is no accompanying letter in (FKS).

11. Kendall to Schaefer, 9 May 1953 (FKS).

12. Hartshorne to Gustav Bergmann, 28 March 1963 (RH).

13. S. B. Jones to Ullman, 7 October 1953 (ELU).

14. Hartshorne to Kendall, 29 October 1953 (RH).

15. Hartshorne to Kendall, 6 November 1953 (RH).

16. Kendall to Hartshorne, 31 November 1953 (RH).

17. Hartshorne to Kendall, 8 December 1953 (RH).

18. Kendall to Hartshorne, 12 December 1953 (RH).

19. Hartshorne to Kendall, 18 December 1953 and 31 December 1953 (RH).

20. Hartshorne to S. B. Jones, copy to Kollmorgen; Hartshorne to Kollmorgen, Ullman, and Jones, 4 April 1955 (RH).

21. This concept of dichotomy is much older. Siddall (1959) claimed that it could be traced through the writings of Leibnitz, Goethe and Schiller, Wundt, Spranger and Fröbel.

22. "Neither on this question, nor any other matter, did Hettner or I use Kant's statement as a basis on which to deduce anything about geography. Rather it served simply as apparently strong confirmation for the theory that Hettner had constructed—without then being aware of Kant's statement—a theory to justify logically the basic conclusion about geography which he had reached first from empirical study of its development" (Hartshorne 1976).

23. Hettner to Hartshorne, 12 May 1940, and 6 February 1936 (RH). In the 1936 letter, Hettner writes, "In all principal points . . . we agree indeed overall. . . ." This letter was written after Hartshorne's "Recent Developments in Political Geography" appeared in the *American Political Science Review*, 1935, 29:785–804, 943–66.

24. Whittlesey to Broek, 2 April 1954 (DSW).

25. Kendall to Kollmorgen, 18 February 1955 (RH).

26. Hartshorne to Kollmorgen, 4 April 1955 (RH).

27. William L. Garrison to Ullman, 12 October 1955 (RH).

28. Ullman to Martin Stearns, 22 July 1966 (ELU).

29. Mill's model was adopted by numerous authors, including J. Arthur Thomson, J. Russell Smith, O. D. von Engeln, and T. G. Taylor. In the U.S., the model was best known resultant to Fenneman 1919.

30. Davis to Mill, 4 May 1896 (GJM).

31. This venture into geography as social physics was, arguably, anticipated by Henry C. Carey, 1793–1879 (McKinney 1968).

32. It is worth noting at this point that Hartshorne majored in mathematics as an undergraduate at Princeton University and even began graduate study in mathematics. He was one of the earliest American geographers to determine the factors that led to concentration of the iron and steel industry in particular areas (Hartshorne 1928), which "developed out of this very elementary use of statistics" (Dow 1972, 2).

33. Bunge to Hartshorne, dated "Tuesday." Hartshorne estimated the date ca. 1 June 1959 (RH).

34. Kendall to Schaefer, quoting Ullman, 14 April 1953 (FKS).

35. Schaefer to Kendall, undated, ca. 5 May 1953, 3 (FKS).

36. McCarty to Martin, 10 June 1978 (GJM).

37. See also McCarty to Martin, 15 January 1984 (GJM).

38. George Vuicich to Martin, 15 May 1989 (GJM).

39. In the acknowledgements to his book, *Philosophy of Science* (University of Wisconsin Press, 1957), Bergmann wrote that he had been awarded a temporary research professorship for the first semester, 1954–55, which enabled him to complete the book.

40. Schaefer seems to have felt harassed in Iowa City. He claimed to have been under surveillance by the FBI, an experience which surely would have been stressful in view of his difficulties in Germany.

41. John C. Hook to Bunge, 22 September 1961 (FKS).

42. Notes in Schaefer's diary, 10 October–4 December 1952, suggest that he could hardly have been happy; they refer to McCarty as "the swine": "The swine influenced several people against me. . . . The swine and the others are just provincial, have never left the country. . . . The article has been sent off. What will the swine say?"

43. McCarty to Martin, 10 June 1978 (GJM).

44. Rushton to Joel Horowitz, 13 December 1988, Memorandum, "Harold McCarty, His View on Geography and the Development of the Department of Geography at Iowa," p. 1.

45. Schaefer to Hartshorne, 17 May 1950 (RH).

46. Hartshorne to William Warntz, 25 March 1969 (RH).

47. In retrospect, there is evidence that a number of geographers had come to regard *The Nature* as orthodoxy. But it was not Hartshorne who so established the book: geographers were free to create their own distance. See Klimm 1947; Marshall 1982.

48. Bunge to Andrew Clark, undated, ca. early 1976. Clark had died by that time, but copies were apparently sent to each person mentioned in the letter (RH).

49. Hartshorne to Bunge, 21 February 1976 (RH).

50. Hartshorne to Graves, 16 April 1982 (RH).

51. Hartshorne to Allen K. Philbrick, 23 May 1979 (RH).

# References

**Aay, Henry.** 1978. Conceptual change and the growth of geographic knowledge: A critical appraisal of the historiography of geography. Ph.D. dissertation, Clark University.

**Ackerman, Edward.** 1945. Geographic training, wartime research, and immediate professional objectives. *Annals of the Association of American Geographers* 35:121–43.

———. 1963. Where is a research frontier? *Annals of the Association of American Geographers* 53:429–40.

**Ambrose, Peter.** 1969. *Analytical human geography.* Harlow, Eng.: Longmans.

**Bowman, Isaiah.** 1934. *Geography in relation to the social sciences.* New York: Charles Scribner.

**Bunge, William W.** 1962. *Theoretical geography.* Lund Studies in Geography, Ser. C, General and Mathematical Geography 1:viii.

———. 1966. *Appendix to theoretical geography.* Lund Studies in Geography, Ser. C, General and Mathematical Geography 6:203–89.

———. 1968. *Fred K. Schaefer and the science of geography.* Harvard Papers in Theoretical Geography, Special Papers Series A.

———. 1979. Perspective on theoretical geography. *Annals of the Association of American Geographers* 69:173.

**Claval, Paul.** 1967. Qu'est-ce que la Géographie? *Geographical Journal* 133(1):33–39.

**Davis, William Morris.** 1900. The physical geography of the lands. *Popular Science Monthly* 57:157–70.

———. 1903, 1904. A scheme of geography. *Geographical Journal* 22:413–23 and 3(1): 20–31.

———. 1932. A retrospect of geography. *Annals of the Association of American Geographers* 22:211–30.

**Dow, Maynard Weston.** 1971a. Clyde Kohn interviewed by John Fraser Hart. *Geographers on Film*, 19 April.

————. 1971b. Harold H. McCarty interviewed by M. W. Dow. *Geographers on Film*, 14 September.

————. 1972. Richard Hartshorne interviewed by Preston E. James. *Geographers on Film*, 19 April.

————. 1976. William Bunge interviewed by Donald G. Janelle. *Geographers on Film*, 3 November.

————. 1986. Richard Hartshorne interviewed by M. W. Dow. *Geographers on Film*, 6 May.

————. 1989. Remarks of David R. Stoddart. *Geographers on Film*, 20 March.

**Dryer, C. R.** 1920. Genetic geography: The development of the geographic sense and concept. *Annals of the Association of American Geographers* 10:3–16.

**Fenneman, N. M.** 1919. The circumference of geography. *Annals of the Association of American Geographers* 9:3–12.

***Geographical Journal (The).*** 1984. 150:429.

**Gilbert, Grove K.** 1914. *The transportation of debris by running water*. U.S. Geological Survey Professional Paper 86. Washington: Government Printing Office.

**Gregory, Derek.** 1978. *Ideology, science, and human geography*. London: Hutchinson.

**Graves, Norman.** 1981. Can geographical studies be subsumed under one paradigm or are a plurality of paradigms inevitable? *Terra* 93(3):86.

————. 1984. *Terra* 96(3):229.

**Hartshorne, Richard.** 1928. Location factors in the iron and steel industry. *Economic Geography* 4:241–52.

————. 1939. The Nature of geography: A critical survey of current thought in the light of the past. Lancaster, PA: Association of American Geographers.

————. 1954. Comment on "Exceptionalism in geography." *Annals of the Association of American Geographers* 44:108–09.

————. 1955. "Exceptionalism in geography" re-examined. *Annals of the Association of American Geographers* 45:205–44.

————. 1958. The concept of geography as a science of space, from Kant and Humboldt to Hettner. *Annals of the Association of American Geographers* 48:97–108.

————. 1959. *Perspective on the nature of geography*. Chicago: Rand McNally.

————. 1969. Qu'est-ce que la géographie? *Geographical Journal* 135(2):323–24.

————. 1976. Reminiscence—scientism and scholarship in geography. Invited paper, AAG annual meeting.

————. 1979. Notes toward a bibliobiography of *The Nature of Geography*. *Annals of the Association of American Geographers* 69:63–76.

————. 1983. "Exceptionalism in geography"—thirty years after Schaefer. Invited paper, West Lakes AAG Symposium, November, Iowa City, IA.

————. 1988. Hettner's exceptionalism—fact or fiction? *History of Geography Journal* 6: 1–4.

**Hudson, John C.** 1979. Editorial statement. *Annals of the Association of American Geographers* 69:128.

**Johnston, R. J.** 1987. *Geography and geographers: Anglo-American human geography since 1945*. London: Edward Arnold.

**Kendall, Maurice G.** 1939. The geographical distribution of crop productivity in England. *Journal of the Royal Statistical Society* 102:21–62.

**Kenzer, Martin S.** 1988. Musings on the history of geography and the AAG directory. *History of Geography Journal* 6:21–25.

**Khan, Mumtaz.** 1984. Schaeferomania: A fallacy in geographic academia (GJM).

**King, Leslie J.** 1979. Areal associations and regressions. *Annals of the Association of American Geographers* 69:127–28.

————. 1988. H. H. McCarty, 1901–1987. *Annals of the Association of American Geographers* 78:551–55.

**Klimm, Lester E.** 1947. "The Nature of Geography": A commentary on the second printing. *The Geographical Review* 37:486–90.

**Leighly, John.** 1937. Some comments on contemporary geographic method. *Annals of the Association of American Geographers* 27:125–41.

**Marshall, John U.** 1982. Geography and critical rationalism. In *Rethinking geographical inquiry,* ed. J. David Wood. Geographical Monographs, Atkinson College, Department of Geography.

**McCarty, H. H.** 1979. The Cornbelt connection: Geography at Iowa. *Annals of the Association of American Geographers* 69:121–24.

**McKinney, W. M.** 1968. Carey, Spencer, and modern geography. *The Professional Geographer* 20:103–06.

**Mill, Hugh Robert.** 1892. The realm of nature: An outline of physiography, part 1. New York: Charles Scribner's Sons.

**Morrill, Richard.** 1983. Recollections of the "Quantitative Revolution's" early years: The University of Washington 1955–1965. In *Recollections of a revolution: Geography as spatial science,* ed. Mark Billinge, Derek Gregory, and Ron Martin, pp. 57–73. London: Macmillan.

**Mullins, L. S.** 1973. New periodicals of geographical interest. *The Geographical Review* 63:571.

**Myres, John L.** 1940. Review of *The Nature of Geography. The Geographical Journal* 95:398–99.

**Parkins, A. E.** 1934. The geography of American geographers. *Journal of Geography* 33:221–30.

**Rose, John K.** 1936. Corn yield and climate in the Corn Belt. *The Geographical Review* 26:88–102.

**Rushton, Gerard.** 1984. Schaefer and the influence of spatial arrangement on social and economic behavior patterns. *Geographical Perspectives,* Fall 54:1–6.

**Sauer, Carl O.** 1941. Foreword to historical geography. *Annals of the Association of American Geographers* 31:1–24.

**Schaefer, Fred K.** 1947. Abstract: Geographical aspects of planning in the USSR. *Annals of the Association of American Geographers* 37:57–58.

———. 1953. Exceptionalism in geography: A methodological examination. *Annals of the Association of American Geographers* 43:226–49.

**Schmitthenner, Heinrich.** 1941. Alfred Hettner. *Geographische Zeitschrift* 49:461.

**Siddall, William R.** 1959. Idiographic and nomothetic geography: The application of some ideas in the philosophy of history and science to geographic methodology. Ph.D. thesis, University of Washington.

**Tower, W. S.** 1910. Scientific geography: The relation of its contents to its subdivisions. *Bulletin of the American Geographical Society* 42:801–25.

**Van Cleef, E.** 1952. Areal differentiation and the "science" of geography. *Science* 115:13 June, pp. 654–55.

———. 1955. Must geographers apologize? *Annals of the Association of American Geographers* 45:105–08.

**Warntz, William.** 1968. Preface to *Fred K. Schaefer and the Science of Geography,* by William W. Bunge. Harvard Papers in Theoretical Geography, Special Papers Series A.

**Whittlesey, Derwent S.** 1945. The horizon of geography. *Annals of the Association of American Geographers* 35:1–36.

**Windelband, Wilhelm.** 1894. *Geschichte und Naturwissenschaft.* Strassburg: Strassburg Universität.

**Wright, John K.** 1941. Review. *Isis* 33:298–300.

# Geography as Museum: Private History and Conservative Idealism in *The Nature of Geography*

NEIL SMITH

Department of Geography, Rutgers University, New Brunswick, NJ 08903

> "Kant's philosophy, which was a condition of the Western mind for a century and a half and which went through a significant revival at the turn of the century, might well cast more light on the nature of geography than a focus on Kant's specific definition of the discipline."
>
> Vincent Berdoulay (1978, 88)

Arguably, the most influential book in twentieth-century English-speaking geography, Richard Hartshorne's *The Nature of Geography* was embraced almost as a holy text by one generation, utterly spurned by another, and is now a dim historical curiosity for yet another. The substance of the text, now selectively siphoned into the disciplinary purview, is today largely abandoned to its historical context. Yet a reconsideration of this work from beyond the epochal curtains of successive positivist and social theory "moves" in geography reveals considerable insights not only about the recent history of geographical inquiry, but about the present predicament.

With *The Nature*, Hartshorne sought to reveal a logical rationale for geography but succeeded only insofar as he simultaneously sanctified an incipient disciplinary isolationism from which human geography in particular is only now emerging. Ironically, with the reconnection of human geography to the intellectual mainstream in the last decade—the "reassertion of space in critical social theory" as Soja (1989) puts it—some of the salient themes of Hartshorne's tome are resurfacing. It would be reductive incongruity to see in recent epistemological shifts—the empirical turn, the selective rejection of theory, the renewed celebration of the unique at the expense of the general—a simple return to Hartshorne's neo-Kantian conservatism of fifty years ago. But the resurfacing of these questions does suggest that whatever the ferocity of critique aimed at Hartshorne in the 1950s and 1960s, the issues themselves may not have been resolved in the collective disciplinary memory. As the conservatism of the present clearly affects the contours of newly fashionable academic ideas, it could hardly be remiss to reassess the major contribution of the most prominent conservative intellectual in twentieth century geography.

Writing in 1938 and 1939, Hartshorne was motivated by several intellectual and political concerns. Given his didactic habit of confronting opponents obliquely and often dismissively, these targets of *The Nature* are nowhere grouped in a single list, but would include the following. First, there was a certain chaos of competing ideas and directions in interwar American geography and, for Hartshorne, this mitigated against the general good of a unified discipline. Related to this, there was insufficient awareness of and respect for the established traditions of geography. Third, the disavowal of environmental determinism had failed to re-establish an appropriate conceptual foundation internally or to arbitrate the position of geography externally vis-à-vis other fields. Finally, a variety of geographical concepts were being reinterpreted in support of clearly partisan national claims, most obviously but not exclusively in Nazi Germany. The scientific ambition of Hartshorne's project is not always appreciated. He sought no less than to rescue geography from these various deviations, as he saw them, and to establish a defensible "scientific" foundation for twentieth-century geographic inquiry.

The central thesis of this paper is that whatever the intent, Hartshorne's own reconstruction committed geography to a museum-like existence. Its concepts were unearthed and reconstituted from a highly selective reading of the disciplinary history, polished up and showcased as intellectual objets d'art. The museum perimeter was jealously fenced by a ring of conceptual distinctions that kept geographers in and effectively discouraged would-be intruders. Hartshorne is undoubtedly the major single inspiration for Kantian and neo-Kantian ideas in twentieth-century English-speaking geography, and with a sympathetic reconsideration of his work now returning to the disciplinary agenda (Hart 1982)—and indeed a broader recrudescence of neo-Kantian perspectives in critical social theory—it is important to explore critically the philosophical bases and effects of his work in historical context.

If the work of the German geographer Alfred Hettner provided the immediate source for much of Hartshorne's methodological treatise, the philosophical perspective owed ultimately to the critical philosophy of the German idealist Immanuel Kant (1724–1804). Hettner himself was much influenced both by Kant and by the end-of-century resurgence of a neo-Kantian tradition in German philosophy. In this essay, I want to examine the Kantian connection insofar as it enables us to peel away some of the philosophical layers of Hartshorne's approach. His Kantian roots have already been noted (Schaefer 1953; Harvey 1969; May 1970; Livingstone and Harrison 1981) and subjected to some philosophical interrogation, but I want here to link such a conceptual discussion with the practical consequences of Hartshorne's epistemology for the discipline. This will involve a discussion of Kant as well as the neo-Kantian revival. Hartshorne's treatment of the central concepts of region and landscape will be examined in order to illustrate the implications of his neo-Kantian framework.

Second, I want to consider the connection between philosophy and history at the heart of *The Nature*. The generally accepted description of Hartshorne as historicist is too simple; he presents a continuist private disciplinary history that is highly conservative, but he is actively anti-historical methodologically. History is the means by which he establishes a highly selective geographical canon. Where other branches of geographical inquiry have today selectively outlived the Hartshornian orthodoxy, the emergence of a critical and contextual history of geography is only now beginning to escape the internalist and idealist focus that Hartshorne fostered.[1]

We begin by setting Hartshorne's own work in historical context.

## The Historical Setting

The contemporary disciplinary structure of academic knowledge dates to the late nineteenth century. There had, of course, been earlier conceptual demarcations of the academic terrain, but in only a few decades at the end of last century, the academic division of labor was institutionalized, first in Europe, then elsewhere. The rapid expansion of knowledge, the professionalization of discrete specialties, the increasing obligation to tie knowledge to distinct social uses, and the politically volatile implications of a unified, integrated, and often-critical social science ("political economy") all encouraged the administrative specialization and division of academic labor into readily definable disciplines or fields. Each had a distinguishable purpose, focus and/or object of study. As befitted the New World of the late nineteenth century, where economic and geographic expansion primarily involved an assault on the physical environment, the incipient geographical tradition in the United States was heavily physical in orientation. In this context, an emerging geography sought to separate itself from geology by focusing on the earth's surface and on the relationship between physical and human features in the landscape. Yet with the increasing spatial, social and political agnosticism of political economy marking the transition to neo-classical economics (Marshall 1890), the scramble for intellectual turf in what would become the social sciences was especially intense.

The promise and appeal of environmental determinism—"geography's entry into modern science," as Peet (1985, 310) observes—at a time when the industrial transformation of the landscape was increasingly manifest, can only be understood in this context. In addition to explaining the forms and processes of the natural landscape, geographers sought to root themselves in the robust scientific lexicon of the physical and biological sciences as a means of explaining the rapidly changing vistas of the human landscape. This causal link between physical and human environments, the appeal to a pervasive naturalism, comprised the flagstone of geography's claim to a niche in the evolving division of scientific labor.

It was, of course, an abortive move. Environmental determinism was quickly discredited less by its dubious social and political implications than by its untenability amidst the maelstrom of social change that marked the *fin de siècle*. It was indeed the end of an era, a restructuring in which, as was once noted of an earlier transformation, "all that is solid melts into air" (Marx and Engels 1955, 13; see also Berman 1982). Attempting to explain a solid world that was melting in front of their eyes, environmental determinists were caught looking backward to the nineteenth century rather than forward to the twentieth and were quickly metamorphosed into intellectual stone. Whatever its ideological utility in terms of imperial expansion and domestic oppression (Peet 1985), environmental determinism was an obsolete technology for comprehending and manipulating the dramatic social and geographical changes of the time. Bowman was only the most explicit of a new generation of geographers in remarking that World War I, the Paris Peace Conference and the political dramas of the postwar world had expunged all vestiges of environmental determinism from his geographical *Weltanschauung*.[2] The initiative in U.S. geography passed decisively if not exclusively to an unprepared human side of the field.

The geographical dimensions of the *fin de siècle* were profoundly important from an economic, political and historical perspective. If widely perceived at the time, the recognition of the centrality of geography in this historical shift waned, to be reawakened only in recent years. The British imperialist Cecil Rhodes (Beer 1898), the German geographer Alexander Supan (1906), and the Russian revolutionary Vladimir Lenin (1917) all shared a common understanding that the era of absolute geographical expansion by collective capital was essentially over, and that only redivision of global territory between competing nation-states and economies was now possible. Thus the maelstrom of change also ushered in a new period, a new stage of capitalist development, the advent of high modernism (Schorske 1980; Kern 1983). Mackinder (1904, 421) may have captured the spirit best in his classic paper, "The Geographical Pivot of History":

> But the opening of the twentieth century is appropriate as the end of a great historic epoch. . . . From the present time forth, in the post-Columbian age, we shall again have to deal with a closed political system, and none the less that it will be one of worldwide scope. Every explosion of social forces, instead of being dissipated in a surrounding circuit of unknown space and barbaric chaos, will be sharply re-echoed from the far side of the globe, and weak elements in the political and economic organism of the world will be shattered in consequence.

Unfortunately, Mackinder did not take up his own challenge; he looked backward rather than forward to elucidate the geographical pivot of history. Nor was the challenge accepted by the discipline as a whole; the wider insight that economic and political expansion would no longer occur through geographical expansion—that there were few if any new worlds to conquer—passed into the lore of the discipline without the profundity of this shift being exposed or explored (Smith 1984b). In the U.S. as in Europe, the emergence of a new political geography in the interwar period (but barely a comparable economic geography) represented a partial recognition of this new salience of space. But it was ill-equipped and showed little propensity to deal with the myriad explosions of social (and economic) forces Mackinder had anticipated, or their global effects. Ultimately the new political geography was insufficiently reconstructive to prevent Sauer (1941, 1–2) from describing this period as "the Great Retreat."

If the new interwar world called for a renewed geography, it was stillborn (Hudson 1977). Geography eschewed modernism. The great retreat was the not-unavoidable offspring of the "great defeat" of environmental determinism; it marked a withdrawal from the competition to establish academic turf, an isolation from the contemporary ferment in social, economic and political theory, and a growing resignation that geography occupy the fragmented interstices between the more dominant social sciences. Increasingly squeezed out, geography rationalized its isolation via an appeal to the discipline's supposedly definitive synthetic role connecting the natural and social sciences—in reality another interstitial locale, if with a fancy name. In practical terms, the period was dominated by "chorography" and chorology (detailed regional descriptions), "apparently in the hope that by-and-by such studies would somehow add up to systematic knowledge" (Sauer 1941, 2). It was into this frustrating intellectual vacuum that Hartshorne launched *The Nature.*

In its seemingly exhaustive historical and philosophical sweep, *The Nature* provided a crucial generation of American geographers (the first trained as geographers) with unprecedented self-justification, precisely when an outward rather than inward-looking perspective could have been so rewarding. Those who read it were indubitably impressed by its scholarly vista and the stellar historical pedigree (from Herodotus to Kant, Humboldt to Hettner) which they as geographers could claim; no other resumé of geography boasted such expansive or searching erudition. The vast majority who did not read it or never actually finished it nonetheless absorbed the awe imparted by such a work. As the dominant statement of the nature of geography, Hartshorne's book was not effectively challenged for nearly a decade and a half, despite its partisan outlook, and it remained the hegemonic philosophical perspective for at least a further decade.

The reveille of antimodernism, *The Nature* massaged the discipline's internal defensiveness; far from reawakening geographical inquiry, the holy text led the faithful into the wilderness, encouraging further retrenchment, deepening rather than ameliorating the great retreat. As well as a philosophical justification, Hartshorne offered geography a distinguished heritage, but at the expense of any serious entanglement with contemporary intellectual trends. Alongside contemporary efforts to situate the discipline (e.g., Bowman 1934), it was a grandiose achievement without intellectual equal. But it made of geography an elite museum wherein the only conceivable future was given in the dead scriptures of the past. Freud and Weber, Keynes and Radcliffe-Brown were exciting the surrounding disciplines; Gramsci and Lukacs, Marcuse and Lefebvre were struggling toward a more critical social theory with incipient spatial implications. Yet *The Nature* sealed geography in from these emerging discourses, presumably finding them irrelevant. The genuine breadth of Hartshorne's vista was entirely in the past and by its very success internally, it functioned to propel geography further inward, codifying its isolation.

# The Kantian Compages

If Kant initiated the modern German idealist tradition in philosophy, he did so in reaction to the cul-de-sac of eighteenth-century British empiricism. Hume, in particular, derived an agnosticism about causation, resigning himself to the possibility of recognizing empirical regularities but precluding the discernment of causes; just because the sun rose yesterday and in many thousands of yesterdays before offers no certainty that it will rise tomorrow. One of Kant's central contributions to philosophy was the attempt to redress this rupture of theory and empirical observation in which the empirical received priority. He retained epistemological critique as the appropriate mode of analysis, and sought to establish a logic of how we know and construct the world. The critique of concepts and of ways of knowing were means to that end.

Writing prior to the modern academic division of labor, Kant's work spans an extensive range of fields: epistemology and cosmology, logic and ethics, history and natural history, religion and science, education and race, law and aesthetics. Kant has been claimed as a significant inspiration for an equally diverse range of contemporary philosophical traditions from positivism to humanism to marxism. Our concern here is primarily with Kant's epistemology, itself a sufficiently complex, diverse, evolving and sometimes inconsistent literature, that it would be foolhardy and diversionary to attempt a comprehensive summary. Instead it will be sufficient to rehearse a pivotal and familiar plank of Kant's epistemology. This argument is summarized at the beginning of *The Critique of Pure Reason*, arguably Kant's most important work and certainly the most influential in the elaboration of his critical philosophy. It is also an argument that is mirrored in Hartshorne's methodology, and this, rather than any pretense at comprehensive summary, explains its importance here.

Kant proposed that indeed there was a real world comprised of "things-in-themselves," but their transmutation into thought was a complex process. From one side, the thing-in-itself engenders in us a sensation and from the other side, human sensibility and intuition create order out of these sensations. The sensation and its ordering combine in what Kant called the "phenomenon"; the phenomenon intercedes between the thing-in-itself and conceptual discourse, but is a product of them both (Kant 1919 ed., 15–17). The Kantian dialectic thereby attempts to rescue the unity of Subject and Object from the empiricist and positivist resignation to duality; Subject and Object commingle, are interspliced, in the phenomenon, yet are at the same time distinct.

From the vantage point of the twentieth century, the central critique of this approach is well rehearsed. The thing-in-itself remains inaccessible to conceptual manipulation, and there is no possible check that the perceived "phenomenon" has anything to do with the world comprised of things-in-themselves. Insofar as Kant retrieves for knowledge a theoretical or intuitive component, the knower, not the known, becomes the major contributor. Kant's practical concern is to establish a logical system of conceptual critique to which he gives the name "transcendental idealism" (1919 ed., 400); but this idealism, privileging as it does the Subject, has the unintended consequence of reintroducing an alternative dualism between Subject and Object, concept and reality: "All thought . . . must, directly or indirectly, go back to intuitions (*Anschauungen*), i.e., to our sensibility" (Kant 1919 ed., 15).

Among his voluminous discussions of space (Garnett 1939), Kant provides a consistent conceptualization: "We maintain the empirical reality of space so far as every possible external experience is concerned, but at the same time its transcendental ideality" (Kant 1919 ed., 22). Yet he also saw space and time as exceptional. Space and time were not simple categories of experience for Kant but rather forms of intuition; they were a priori, "two pure forms" of intuition, pre-given not in external nature but in the basic human ability to perceive, "a property of our mind" (Kant 1919 ed., 17, 18). As Bertrand Russell (1945, 713) once put it referring to Kant, "we carry . . . about with us" space and time *as* pure intuition. These pure forms of intuition automatically order our empirical experience.

As part of his epistemological critique, Kant was already concerned to derive a logical division of different fields of knowledge. Consistent with his larger project, this early attempt to map the academic division of labor was first and foremost a question of conceptual rigor and consistency, an effort at ordering knowledge as a means to order real world events. He proposed two separate classification systems. On the one hand, knowledge could be classified logically or systematically according to perceived differences in phenomena—conceptually parallel to the contemporary notion that different fields have different objects of study. On the other hand, given the importance of space and time as a priori modes of intuition, it is also important to recognize a certain deep-seated priority to spatial and temporal knowledge of the world as a whole. This task fell to the fields of geography and history respectively, for Kant:

> Description according to time is history, that according to space is geography. . . . History differs from geography only in the consideration of time and area (*Raum*). The former is a report of phenomena that follow one another (*nacheinander*) and has reference to time. The latter is a report of phenomena beside each other (*nebeneinander*) in space. History is a narrative, geography a description. . . .
>
> Geography and history fill up the entire circumference of our perceptions: geography that of space, history that of time (Kant 1923 ed., quoted in Hartshorne 1961 ed., 135).

Hartshorne shares two central planks of the Kantian epistemology. In his internal discussion of the nature of geography, he employs a broadly Kantian critique of concepts as the appropriate epistemological procedure. In his external treatment of geography's "position among the sciences" (367),[3] the very statement of the question no less than the form of the answer again parallels Kant. We shall briefly examine each of these issues.

In his interrogation of the nature of geography, Hartshorne sought to establish a logically consistent philosophical foundation for the discipline. This involved a systematic consideration of the central concepts of geography, critique of potentially contradictory or illogical meanings, clarification of confusions, and establishment of more or less technical definitions for basic concepts: "The fundamental requirements of logical reasoning in any science demand that the basic terms of methodological discussion must be precisely defined and the definitions adhered to in the discussions" (158). Alerting us to the anticipated difficulty of the project, he warns in connection with his discussion of regions that "however tortuous the path others have laid for us to follow, we will clear our way through to daylight" (252). To the unprepared reader, the painstaking semantic dissection of "region," "science" and "landscape," the question whether geography studies only "things perceived by the senses," the investigation whether regions are "unified wholes" or even "real objects," and so forth—such extended and repetitive discourses (repetitive in the interests of logical completeness) must indeed have seemed tortuous, pedantic, scholastic in the extreme. And yet this treatment displayed a certain consistency.

In *The Nature*, Hartshorne works through the concepts of geography in systematic fashion, imposing logical order. Akin to Kant's first classification of knowledge according to differences in phenomena, Hartshorne conceives knowledge more as geographical terrain than historical process. He envisages vast continents of knowledge subdivided neatly into discrete subjects, separated by sharp national boundaries. A homologous neatness of subdivision pertains internally between specialties, but also with regard to certain basic concepts; these may be conceived as taking the place of key political institutions in the nation-state. Much as the traditional political scientist might survey these institutions one by one in introductory courses on government, establishing their principle features and core meanings, and demarcating their jurisdictional boundaries, he sorts through the concepts of geography to establish a neat conceptual landscape, devoid of gaps and overlaps.[4] Hartshorne's is a quintessentially constitutional methodology; it treats geography as a republic rather than a democracy.

The idealism of this approach does not remain abstract, inhering simply in the strategy of epistemological critique, but has very practical consequences. In the first place, implicit in *The Nature* is the belief that the practice of geography can be kept on the right track, or returned there, via logical conceptual renovation; once the conceptual basis of the discipline is consistently and logically revealed, only fools or knaves will depart from the proper course. That ideas guide practice is indisputable, but as we shall see below in greater detail, Hartshorne provides little or no room for changing practices to alter the conceptual foundation of geography. This has had a particularly deleterious effect on the discipline. The division of the sciences is deemed to result from some overriding conceptual logic, not the action of the sciences or scientists themselves or their engagement with the real world. As Kant put it in the introduction to "Physische Geographie," we must divide knowledge "into definite disciplines even before we obtain the knowledge itself" (Kant 1923, trans. in May 1970, 255). (Extrapolating this argument to the present predicament, if geography is not given its due or is not in command of all its rightful intellectual terrain, this is the result merely of illogic, which of course will be corrected once carefully explained to the academic powers that be.)

In the isolation of concepts from the active world to which they refer, Hartshorne's method duplicates Kant's philosophy. We shall illustrate these conclusions by direct reference to Hartshorne's conception of the region and his disavowal of "landscape," but in the meantime, it is necessary to consider the second major debt to Kant, namely the external vision of geography's position among the sciences.

As forms of "pure intuition" for Kant, space and time provide privileged intuitive means for ordering experience. Hartshorne (135) enthusiastically adopted Kant's claim that "geography and history fill up the entire circumference of our perceptions," thus distinguishing them from the purely systematic sciences. It was this stance that drew Schaefer's critique of exceptionalism. For Schaefer, exceptionalism claimed too much; geography was not separate from and above the other sciences. Schaefer sought to explain exceptionalism historically as a "hangover from the time when there were no social sciences and not much natural science, and when such quaint and encyclopaedic endeavors as natural history and cosmology still occupied their place" (Schaefer 1953, 231). Without miring ourselves in the Hartshorne-Schaefer controversy, we might concede that whatever complaints Hartshorne (1955) would voice about the mores of debate and despite a recent contention that "'exceptionalism' is a misnomer" (1988, 4), the essence of Schaefer's critique is indisputable. Yet today the exceptionalist argument still lingers, especially when combined with the hoary old notion that geography is essentially synthetic or integrative of more systematic researches—a "mother of sciences" (373).

Kant himself seems to have been rather more ambivalent about the value of geography than Hartshorne's appropriation of his work would suggest. In the first place he treated geography as a "propaedeutic" (Kant 1923 ed., 157), meaning a preliminary survey in preparation for more advanced inquiry; geography was an essential foundation for higher study. As such Kant's geography largely comprised empirical generalizations and categorizations of objects and events in the landscape, as indeed was pointed out by the German geographer Georg Gerland (1905, 508–09). In a 1757 "Proposal for and Announcement of a College of Physical Geography," Kant explained how he proposed to view physical geography:

> not with that completeness and philosophical exactitude in each part which is a matter for physics and natural history, but with the rational curiosity of a traveler who everywhere seeks out what is noteworthy, peculiar, and beautiful, collates his collection of observations, and reflects on its design (quoted in Cassirer 1981, 52).

That philosophers have tended to neglect Kant's geographical writings may be in part the product of their own disciplinary snobbishness (May 1970) and in part due to a pervasive philosophical historicism (Soja 1989). But it may also reflect something of Kant's own priorities and even the worth of his substantive geographical ideas (Schaefer 1953). Whatever its foundational value, the subject played virtually no role in his philosophical investigations; it seems not to have been mentioned at all in *The Critique of Pure Reason* and is only fitfully acknowledged among Kant scholars. Thus it may not be too outlandish to suggest that Kant's involvement with geography represented a philosophical obligation as much as any scientific commitment. As Cassirer (1981, 52–53) observes, "it is in general an ideal of comprehensive human wisdom at which Kant aims in his own growth as well as his teaching. . . . the lectures on physical geography . . . pursued this goal." The teaching of geography, Kant felt, would facilitate a "unity of knowledge, without which all learning is only piece-meal" (May 1970, 68).

That geography for Kant was obligatory propaedeutic need not be taken as demeaning the importance of an academic subject still in the throes of formation. At the same time, however, we ought not to mistake obligation for virtue as Hartshorne seems to do in declaring that "our field received for many years the attention of one of the great masters of logical thought" (134). Hartshorne was certainly aware of the propaedeutic nature of geography for Kant and of Gerland's reservations (38). But with a certain disciplinary anxiety for recognition and respectability, and with not a little blindness to historical differences, Hartshorne, in his encomium for Kant, suspended all criticism. The intellectual respect to be engendered by this short-cut to distinguished philosophical pedigree would not, alas, materialize.

## The Neo-Kantian Revival

That Hartshorne, along with so many other twentieth-century thinkers, was so clearly influenced by the ideas of the eighteenth-century philosopher, who undergirded so much of the last two centuries of Western bourgeois thought, should hardly be controversial. Yet he consistently deflects the suggestion of Kant's influence on his philosophy and methodology (Hartshorne 1955, 219–20 fn.; 1972, 78). In 1939, at the time *The Nature* was written, he had read secondary sources on Kant, but his direct familiarity with the Königsberg philosopher was limited to the latter's discussions of geography. Only in 1970 did he tackle Kant's philosophical work (personal communication, 20 May 1989).

In addition to his limited primary and secondary knowledge of Kant's philosophical work, Hartshorne absorbed (primarily via Hettner) much of the philosophical inspiration behind *The Nature* from the neo-Kantian resurgence that coursed through continental European thought, achieving its apotheosis in the earliest years of this century. If Kant's own work was majestically diverse and complex, the "variegated spectrum of neo-Kantianism" (Willey 1978, 37) was even more so. Essentially conservative in seeking to re-center the individual and normative human values in the construction of the world, as against the received power of history and science, neo-Kantianism was, by precisely the same token, a progressive revolt against absolutism and a cornerstone of a new political liberalism. At times it embodied a direct critique of German marxism, yet there were also neo-Kantian socialists; and an equally blunt rebuttal of Hegel did not preclude a virulent neo-Kantian nationalism.

In his sympathetic treatment of the Kant revival, Thomas Willey finds at least "seven different species of neo-Kantianism," but discerns among these "four common Kantian assumptions":

1. they employ "the transcendental method";
2. they are "conceptualists," implying a belief in "the capacity of reason 'for constructing a whole from its parts'";
3. their "epistemologies are idealist";
4. they hold that to "understand Kant is to go beyond him."

As regards this last assumption, Willey argues that the neo-Kantians "all reject the unknowable ground of experience, the notorious thing-in-itself" (1978, 37).

Hettner, of course, borrowing from the philosophical work of Wilhelm Windelband and his student Heinrich Rickert, the two principals of the Baden school of neo-Kantian philosophy, was the main conduit for such ideas into Hartshorne's work (May 1970). Although Hartshorne (1972) again tries to deflect if not deny this connection, Windelband's own 1903 testimony suggests how pervasive the Kant revival in fact was. "The Kantian critique was so generally taught as a point of departure for all philosophical thinking that it influenced many scholars who were not professional philosophers. Kant's position affected almost every aspect of German learning" (quoted in Willey 1978, 131). Hettner, who was trained in philosophy and almost followed a philosophical career (Hartshorne 1958), himself acknowledges the debt (Hettner 1927, 112–14).

What distinguishes Windelband and Rickert in the larger neo-Kantian revival is their pursuit of a logical rationale for history grounded in a reconstructed concept of the historical individual and their related effort at a universal and normative theory of value. It was Windelband who distinguished the emerging historical and social sciences from the natural sciences (which Kant emphasized) and attributed to them different methodologies—the idiographic and nomothetic respectively (Windelband 1980; Rickert 1962, xi–xiii). Whereas the nomothetic sought to explain events by means of general laws, the idiographic sought to understand individual events. Very much following Kant, Windelband's work was thoroughly methodological; he focused on the logic of historical inquiry rather than the actual historical events and social processes that attracted others, including Max Weber, Durkheim, Cassirer and Simmel, who were also influenced by neo-Kantianism. He shared with Rickert a more abstract conceptualism; the analysis of political questions was largely neglected as trivial. "Windelband and Rickert were more typically 'mandarins'" (Willey 1978, 133).

Kant's dilemma of the dualism between concept and reality, the unknowability of the "thing-in-itself," was dubbed by Fichte the *hiatus irrationalis*, i.e., the gap that isolates concrete reality from rational thought. The *hiatus irrationalis* presented a major obstacle to Windelband's and Rickert's attempts to provide a scientific methodology for historical inquiry. Rickert, in particular, attempted to resolve the dualism. It is a complex and involved argument but can be summarized quite succinctly. For Rickert, although concrete reality itself lay beyond the reach of conceptualization, the *individuality* of some concrete reality was eminently conceivable. But this raises a fundamental problem of how we are to decide what constitutes a coherent individual event, experience or object. How do we identify the boundaries that determine the difference between one individual event and another? This problem can only be solved, according to Rickert, by the explicit adjudication of what constitutes a significant "historical individual"; this in turn requires a normative theory of value, which is meant to provide an objective footing for identifying and conceptualizing the historical individual (Oakes 1987).

In the specific analysis of Hartshorne's concepts of landscape and the region, it will be possible to identify similarities but also divergencies with various dimensions of this, perhaps the most conservative, branch of neo-Kantianism. In the meantime, it is important to stress that the neo-Kantian revival was not restricted to its German homeland, nor were Hettner and Hartshorne by any means the only geographers influenced. To cite the most obvious example, Vidal's concept of geographical individuality draws on neo-Kantian treatments of history; his concern for the identity of place sought to balance the integrity of externally experienced data with "Kant's idea that the mind imposes order upon the world" (Berdoulay 1978, 82; see also Livingstone and Harrison 1981, 361). Durkheim's sociology rested on a conspicuously more positivist derivation of Kant. Indeed Berdoulay (1978) depicts the debate between Vidal and Durkheim as not simply a border skirmish between emerging disciplines, but equally a philosophical dispute between differing neo-Kantian visions.

# The Hartshornian Region

*The Nature of Geography* is about the proposition, exposition and defense of a concept—geography. Within the confines of geography, the regional concept plays a central constitutional role. Hartshorne tackles the resurgent "belief" that regions are concrete entities, often associated with the view that the earth's surface comprises a mosaic of landscapes and regions. It seems likely that he is reacting here to the German revival of interest in Ratzel's organic regionalism with the rise of Geopolitik in the 1920s and 1930s, but this is not made explicit. While accepting that "the areal differentiation of the earth's surface is a 'naively given fact'" (251), Hartshorne precludes the application of this a priori principle to regions. In the first place, he cites the difficulty of defining regions or even agreeing upon criteria of definition, concluding that they are actually somewhat arbitrary (252). Second, he rejects any suggestion that regions are organic wholes since, with Albrecht Penck, "an organism is essentially indivisible, whereas any regional unit of the earth's surface can be divided into smaller and smaller units and these into still smaller units" (259). Third, since a regional whole is no more than the sum of its parts, but since some of these parts are either unknown, unknowable or unrelated to other parts, it is impossible to interpret the "region as a whole" in any meaningful way. Finally, even if these difficulties could be overcome, the "problem of drawing the limits" of any particular region (267) is interminable; no single boundary for a region will be universally acceptable or applicable.

Hartshorne's view of regions is quite distinct, but the important point here is his destination. While not denying that regions may be *experienced* as wholes, Hartshorne dismisses this form of "holism" as the result of experience and intuition, and therefore a "psychological phenomenon" more properly dealt with by psychology (276–77). From the point of view of geographers, there is no coherent whole "less than the entire earth itself"; sub-global regions are not "determined in nature or in reality" (283–84), but are essentially "arbitrary" (xii). "The regional entities which we construct . . . are therefore in the full sense mental constructions; they are entities only in our thoughts, even though we find them to be constructions that provide some sort of intelligent basis for organizing our knowledge of reality" (275).

The Kantian *hiatus irrationalis* is perfectly replicated in Hartshorne's vision of the region. No thing-in-itself, the region as concept is even more sharply severed from reality than in Kant's dialectic. For Kant, the "phenomenon" combines intuition with empirical sensation, whereas for Hartshorne the regional concept exists "only in our thoughts." Indeed Hartshorne recognizes this; his wording is, as ever, precise, and he makes no claim that his regional concept, suitably defined, provides more intelligent access to real world geographical events and objects. He claims only that his concept provides *"some sort* of intelligent basis for *organizing* our knowledge of reality" (emphasis added), a knowledge presumably already achieved, although Hartshorne does not explain how.

The finale of this major discussion of regions is surprisingly anti-climactic. For Hartshorne, the "region" is the central organizing concept in geography, and he identifies as a fundamental disciplinary ambition the systematic classification of regions and the identification of different methods for classifying the world into regions, albeit regions that are "not inherent in the world" (362). Yet even this proto-scientific ambition has met with little success. It "has produced no simple system of classification of areas, whose general outline can be recognized on the basis of our present knowledge of the field" (362).

In practice, if not in intent, Hartshorne's severance of regions from the material world nurtured an anti-intellectual relativism in mid-century regional geography. If regional differentiation was to be seen first and foremost as a conceptual question, the variety and criteria of regional classifications of the world were limited only by the imagination of geographers. Who was to say that a division of Germany into economic regions was any more or less valid than its division according to shoe size? Should France be dissected according to geomorphological regions or differing rooftop colors? The qualifier that such regional schemes be chosen with regard to "significance to man" or according to the purpose of research merely begged the question.

This was an extraordinary result. The central concept of geography was effectively denied any material referent by the most influential philosophical text emanating from within the discipline itself. Hartshorne's neo-Kantian idealism justified and indeed promoted the sterility of a regional concept and approach at precisely the time when real-world landscapes were being dramatically restructured by the Depression, World War II, the Cold War and the Pax Americana, postwar economic expansion, suburbanization and the professionalization of urban planning. Burned perhaps by too close a connection with real-world affairs occasioned by environmental determinism, the discipline recoiled from current events despite the dramatic geographical transformations involved. One could hardly tred further from contemporary events than a neo-Kantian discourse on ideal regions. The positivist revolt of the 1950s and 1960s, initiated in part by Schaefer's paper, reacted precisely to this internal sterility and external irrelevance.

The expected postwar expansion in demand for geographical knowledge was real enough. Before the end of the 1940s, Harvard and Columbia Universities had established institutes for Soviet studies leading there and eventually throughout the U.S. to schools of international affairs. But as Sauer and others sensed, even at the time, geography would be passed by in this expansion of area studies (Prunty 1979, 58). The habitual non-involvement of geography and geographers in these enterprises was in no way a virtuous reaction to their nationalist and generally right-wing ideologies, but, at best, resulted from a naive rejection of the relevance of politics per se within geography. It would, of course, be an exaggeration to blame *The Nature* alone for the isolation of geography at mid-century and for the inability of geographers to capitalize on the emergence of area studies. Yet especially in its treatment of the central concept of regions, this work rationalized the incipient disciplinary defensiveness by encouraging geographers in their abstentionism. The great retreat did not end with *The Nature* as Sauer (1941) expected, but was deepened by it.

Hartshorne's concept of the region reveals a further dimension of his thought; it embodies a pure expression of absolute space. Space for Hartshorne is infinitely divisible, an abstract field of experience, a coordinate (along with time) for ordering reality. Events, objects and processes do not constitute space but happen "in space." Regions, as we have already seen, are infinitely divisible into their non-organic parts. They "can be divided into smaller and these into still smaller units. . . ." Having denied any holism of the region, Hartshorne elaborates in a vision that, insofar as it is not tautological, can only be described as astonishing, coming from a geographer:

> Foreign capitalists and engineers may insert factories in a region of primitive subsistence economy, as though a surgeon were to put a backbone in a star-fish. . . .
> The region as a whole does not undergo changes, but only the complex of different regional elements changes with changes in its elements (259–60).

A more expressive evocation of the concept of absolute space could hardly be imagined.

The absolute conception of space is much too ubiquitous for us to expect to trace its lineage in Hartshorne's work. It derives in the modern period from Newton and Descartes as well as Kant who, in one of his arguments to prove that space was an a priori intuition, appealed to the self-evident nature of Euclidean geometry and the infinitude of space (Kant 1919 ed., 19).[5] The sustained Newtonian influence in Kant is suggested by Hartshorne himself (1955, 221 fn.). What is surprising, for a geographer, is the philosophical exclusivity with which spatial absolutism is exercised. Almost a quarter-century after Einstein's highly publicized general theory of relativity, which emerged precisely when absolute geographic expansion was superseded by relative expansion, no alternative to the absolute conception is entertained and no discussion of the geographical implications of the relativity of space is offered. He is either unaware of spatial relativity or else he partitions it off as physics, philosophy or something else beyond geography. Such a discussion, in fact, would not emerge for almost another quarter-century. Hartshorne does concede that "relative location" (not the relative conception of space per se) is a "factor of great importance" (364), but it plays only a trivial role for him. Central as it has been in geography, relative location is relegated to appendectic status in *The Nature*, tacked on at the end of the two chapters on regions (282–84, 364).

In his discussion of regions as wholes, Hartshorne confronts a parallel dilemma to the one Rickert tackled with his concept of the historical individual. The only geographical "whole" Hartshorne has thus far allowed is the globe itself, and yet in proposing that the globe be divided into regions, it is incumbent on him to explain what it is that comprises these regions. If the region cannot reasonably be held to constitute the geographical individual, what does? Regions, he says, are "element complexes," discretely conceived areas made up of distinct combinations of elements (the whole is constructed from its parts). As Hartshorne recognizes, this is not an answer but rather reconstitutes the question at a different scale. How are we to recognize or constitute these basic geographical "elements," or are they also infinitely divisible? What kind of element complexes might reasonably be considered as "wholes?" Hartshorne answers that we can identify "culturo-geographic regions" (354) in terms of land use units. So how are we to identify land uses in an area? This is a question about the scale at which we wish to generalize concerning land use; what is the size and structure of these basic units? Hartshorne resorts to an agrarian essentialism; "the individual farm, plantation or ranch" is rendered the geographical individual:

> The farm, as an organized unit, includes not merely the land and the plants and buildings on it, but also the livestock, tools, methods and intensity of production, and the use of the products. In other words, the farm represents not merely an element-complex—such as is found in areas of wild vegetation—but is a primary Whole. . . . Each and all of the elements listed, whether material fields, buildings or tools, or immaterial methods of production, can be understood in form and function only in terms of the whole farm unit (351).

Thoroughly commensurate with Rickert's pursuit of the historical individual, this discussion of regions and farms is also bound up with Windelband's distinction between the idiographic and nomothetic methods. From our earlier discussion, it should be evident that the concept of the historical individual embodies an inherent tension between uniqueness and generality, and no less is true of Hartshorne's choice of the farm as the geographical individual. This is an important subtlety in Hartshorne's concept of the unique (379–84), often overlooked by Hartshorne's positivist critics and indeed more recent champions of geographical uniqueness. "Farm" is both a generic concept and, at the same time, its instances are absolutely specific, clearly identifiable, unique spaces. Generalization is inherent in the adjudication of uniqueness; it would be entirely inconsistent for Hartshorne to exclude the nomothetic method, no matter how much he in fact privileges the idiographic.

At first glance Hartshorne's geographical solution to this central neo-Kantian dilemma is strikingly arbitrary. He is nowhere clear why the farm is to be privileged over, for instance, the individual field, the county, the valley or the city as the "primary Whole." Nor is it clear why the elements of the *farm* can be understood "only in terms of the whole farm unit" while the *region* as a whole remains unaffected by the insertion or excision of its economic backbone. If his largely unexamined Kantian assumptions have led him into an episte-mological cul-de-sac, Hartshorne's more acute geographical sensibility did not follow, and, in later revisions and discussions, he backs away from this conclu-sion.[6] In the end, he admits a clear epistemological role for generalization but wants to restrict severely its jurisdiction, which he does by identifying a max-imum scale at which generalizations can be taken to represent realities also conceived as unique. The uniqueness of the regional concept is too important in Hartshorne's geography for it to be compromised as a bearer, too, of gen-erality; the geographical individual was to be sought at a higher scale of reso-lution. But why the farm?

If Hartshorne's philosophical idealism enticed him astray from material events and real world affairs, his thought was not completely ungrounded. The logical requirement that he identify sub-global "wholes" out of which to construct "element complexes," and in turn his regions, returns him to earth with a prosaic thud. Nor is he entirely imprisoned in the eighteenth century insofar as "the farm" is his desired "primary Whole." What is striking about Hartshorne's choice of the farm, especially in the light of his prior discussion of capitalist factories inserted into regions, is that the basic units or elements comprising "element-complexes" are humanly constructed absolute spaces. The factory and farm are also geographically defined elements of the means of production, i.e., units defined by the rules of private property ownership. Where Windelband and Rickert recognized the need for a universal, normative theory of values to resolve the *hiatus irrationalis*, Hartshorne resorts to the legal formalism of cap-italist property rights.

With his regional concept, Hartshorne is both too much and too little a Kantian. With the election of the farm as the geographical individual, histori-cally bound social relationships cross the *hiatus irrationalis* to impinge on concept formation. Yet by insisting on the region as a conceptual construct divorced from concrete reality, he has not grasped the insights of the Baden neo-Kantians and remains mired in Kant. Ironically, it was none other than Windelband who penned the neo-Kantian motto: "Kant verstehen heisst über ihn hinausgehen" (to understand Kant is to go beyond him) (Willey 1978, 135).

# The Assassination of Landscape

If his discussion of regions is an effort to construct a usable category, Hartshorne's approach to landscape is wholly destructive. The argument was sufficiently successful in convincing succeeding generations of English language geographers that the notion of landscape has "little or no value as a technical scientific term" (158), that as a result this "single most important word in geographic language" (149) has been largely excluded from serious theoretical discourse almost to the present day. During the writing of *The Nature* in Vienna immediately following the *Anschluss*, Hartshorne had considerable interaction with German geographers, and it is plausible that the uncompromising dismissal of the landscape concept reflected not simply his commitment to conceptual order but also a disavowal of the entanglement of aesthetics and politics espoused in the more nationalist and mystical treatments of *Landschaft* (Banse 1928). Characteristically, there is little hint of this in the text. His argument can be summarized as follows.

The term landscape is inherently confusing, especially as the German "Landschaft," which can refer both to the general appearance of the land and to a distinct area. Such confusion could not provide a logical foundation for geographic inquiry. Second, it is unclear how landscape differs from area; both are composite spatial categories encompassing distinct phenomena and both are divisible. Geographers already have a perfectly acceptable word for specific areas—they are regions. Further, "landscape" refers only to the surface of the land: "the landscape is literally a superficial phenomenon and a field of science that concentrated on it alone would be superficial" (165). Finally, Hartshorne rejects Sauer's conception of landscape as overly narrow with its emphasis on physical objects and form (155) and concludes that the notion of a "natural landscape" is "a theoretical conception that not only does not exist in reality, but never did exist" (173). To employ today's terminology, landscape for Hartshorne was, at the very least, a "chaotic conception" (Sayer 1984). Yet this may reflect as much upon the Kantian roots of the realist notion of "chaotic conception" as on the concept of landscape.

This dismissive argument is the least subtle of Hartshorne's terminological duels with opponents real or imagined, yet it struck a chord among geographers genuinely confused about "landscape." It is a piece of definitional essentialism that not only harkens back to neo-Kantian squabbles, but anticipates the rigid conceptual formalism of positivist thought. It is a thinly disguised polemic. In the multiple meanings of landscape, Hartshorne could see only definitional confusion, not opportunity; the richness of "landscape" eluded him. As Cosgrove (1984, 15) indicates, Hartshorne and other proto-positive thinkers seek the "removal of ambiguity" in the concept of landscape. Where richness of meaning existed for others, it was summarily reduced to a single dimension in preparation for dismissal.

Without at all denying the confusion inherent in the concept of landscape, nor indeed its periodic function as a "vehicle for a reactionary geographical ideology" (Cosgrove 1984, 261), a reassessment would surely be timely and fruitful. What attracted Sauer and others to the concept of landscape was in part its potential to bridge the gulf between space and society, i.e., to diffuse the dualism engendered by the absolute conception of space, although it was hardly put in these terms. In the landscape, spatial extent and form together with material content are mutually imbricated with each other, spliced into a tangible as much as symbolic reality. For Sauer the material dimension of landscapes linked morphology and culture; taking the organic metaphor as a natural reality, the human ecology school in the same period can be seen as seeking a similar reconciliation of space and society (Barrows 1923; Park 1925). The success of the Chicago School derived in part from this ability to combine these two realities that were traditionally held separate. Especially for Sauer, it became possible to think in terms of making historical landscapes, making geographies, although he himself emphasized morphology at the expense of process (Solot 1986).

There is no attempt here to see Sauer, the Chicago School, or the German proponents of *Landschaft* as the true founders of geographical relativity. Sauer was clearly antithetical to social theory and tended, not unlike Hartshorne, to treat history as a sequence of ad hoc events. To the extent that landscapes were made, for Sauer, this was a voluntaristic rather than collective social result (Solot 1986). Nor was the Chicago School immune from reactionary ideology, especially as regards race. If these represented alternatives to the Hartshornian vision, it is also important to point out that in different ways, they too were influenced by the neo-Kantian revival. Himself influenced by contemporary German geographers and philosophers, Sauer mixed an uneasy combination of pragmatism and neo-Kantianism, according to Entrikin (1984, 388). Robert Park of the Chicago School received his Ph.D. under Windelband's supervision and also studied with Hettner (Entrikin 1980).

The potential interlocking of space and society in the concept of landscape was directed by Sauer entirely at the past. But, as with the Chicago School, it was also amenable to comprehending the dramatic transformations that marked the turbulent geography of early twentieth-century capitalism—more so than an approach that insisted on holding space as an etherial conceptual given, inaccessible to the mere social and material world. An exciting potential was available in the concept of landscape, in all its diversity. It remained simply a potential, due in large part to the authority of Hartshorne's conceptual assassination.

It is not difficult to see why "landscape" was so anathematical for Hartshorne. Such a priori muddling of space and society had contributed to the debacle of environmental determinism. But more, in combining events and experiences, objects, processes and space into a single notion, organic or otherwise, it broke all the Kantian rules of logic that were aimed in the opposite direction toward the continual separation and logical classification of different phenomena. Further, insofar as it challenged the assumption of absolute space, it threatened the exceptionalism of geography. Hartshorne's rejection of landscape represents a momentous wrong-turn which not only contributed to geography's arcane isolation but, in an ironic twist, abetted the mid-century decline of regional geography. Probably no other aspect of geography would have benefitted as much from a more sophisticated conception of landscape. With a dialectical conception of geography more rooted today, and the mutual interdependence of social theory and geography increasingly evident, the time has proven ripe for a more serious and long overdue reexamination of landscape that moves beyond narrowly descriptive, aesthetic and idealist confines (Cosgrove 1984, 1985; Cosgrove and Daniels 1988; Cronon 1983; Olwig 1984).

## "Private History" and Historicism

It has become traditional to describe Hartshorne's work as "neo-Kantian historicism." If the roots in Kant would seem to be clear, the argument about historicism is more complex and somewhat more obscure. Hartshorne's own treatment of history is widely replicated today in the "history of geographical thought," but whether it comprises a historicism is less clear. In all likelihood, it was Schaefer who set this pattern of response to Hartshorne, yet the epithet continues to be repeated rather uncritically. According to Schaefer (1953), Hartshorne constructed a linear, evolutionary narrative of events and ideas in geography and attempted to fashion a vision for contemporary investigation that would fit this historical schema. Such historicism, Schaefer added, was antiscientific in spirit, dominated much nineteenth-century German thought, and, through Hettner and eventually Hartshorne, has "powerfully affected" the course of geography.

If somewhat scattered, Schaefer's depiction of Hartshorne's treatment of the history of geography ("the history of geographical thought" in the idealist tradition) is broadly accurate. His mistake lies in the unqualified generalization that Hartshorne's work is historicist. In his response to Schaefer, Hartshorne was generally evasive and defensive, deflecting much of the critique with the retort that there was "no evidence." To the accusation of historicism, however, he is more forthcoming, quoting his own (and Hettner's) warnings about straying too far and crossing the line into history. He even cites an earlier criticism of Hettner's work as "static" (1955, 277 fn.).

Hartshorne is of course correct; *The Nature of Geography* is no champion of history. Rather, Hartshorne admits a certain circumscribed utility for the "genetic concept," but conceives of historical geography so narrowly as to jettison much of this work as beyond geography: "Historical geography, therefore, is not a branch of geography, comparable to economic or political geography. Neither is it the geography of history, nor the history of geography. It is rather another geography, complete in itself, with all its branches" (184–85). A curious conclusion, but it seems to have been little challenged. Carl Sauer, Jan Broek, Ellsworth Huntington, Vernor Finch, Stanley Dodge and Glenn Trewartha are but several of the geographers whom Hartshorne suspects of transgressing the line with history (175–83, 212). Leighly's (1937) heresy that cultural geography might experiment with a historical point of view tempts Hartshorne to the conclusion that "this is the antithesis of geography" (179). "We are forced to distinguish between an historical and geographical point of view, and in order to master the technique of either, we need to keep clearly in mind the distinction between the two" (188). If anything is forcing Hartshorne here, it is the logical application of Kant's distinction between space and time.

Equally, Hartshorne is antithetical to "process." While some of this sentiment is expressed at the expense of geomorphology, he also takes aim at Dodge (1936) who made the mistake of admonishing geographers that regions be viewed not as static but as "becoming" (182–83); the question of development is also essentially historical and therefore non-geographical. Even "the idea of defining geography in terms of relationships" (120) is suspect.[7] Were Hartshorne to specify this latter notion historically, as for example a rejection of the spurious relationships supposed by environmental determinists, it would be unobjectionable, but with a penchant for the universal, he overstates an otherwise plausible idea.

Hartshorne's anti-historical perspective is not difficult to fathom. At one level, he is simply mounting a jealous if principled defense of disciplinary boundaries against the major competitor: space against time. But the epistemological idealism of Kant, as much as his ordering of the sciences, lends itself to Hartshorne's cause. Thus Rorty (1979, 9) summarizes the universal imperative of Kantianism: "traditional Cartesian-Kantian" philosophy of the eighteenth century can be comprehended "as an attempt to escape from history—an attempt to find non-historical conditions of any possible historical development." In his substantive discussions of what geographers do and should do, Hartshorne adhered to this anti-historicism. In justifying his view of geography (1958), he attempted to show that Kant, Humboldt and Hettner shared the same concept of geography, and was adamant that they reached it essentially independently. One implication was that their (and Hartshorne's) concept of geography was given of universal concepts and not the product of historical construction.

Yet as this suggests, despite his general antipathy to history, Hartshorne constructs a very specific disciplinary history on which his entire case is built, and it is to this that Schaefer reacts in his discussion of historicism. Every discipline, Hartshorne says in a footnote, "has its own, one might say, private history" (184 fn.), and it is indeed such a private history that he crafts. Despite the intellectual authority it is made to bear, Hartshorne's is a quite anti-historical history. As in a museum, the intellectual artifacts are arranged in historical sequence, room by room, for the purpose of glorifying the present. Geography is today simply what it has always been: its roots "as a field of study, reach back to Classical Antiquity" (35) and the work of the earliest geographers, Herodotus and Strabo. "This study, since the days of antiquity, they have called geography and the world of knowledge has recognized it under that name" (115). The entire perspective of *The Nature* is to provide philosophical justification for the supposedly continuous tradition deemed to have existed since antiquity. The purpose of establishing the disciplinary heritage is that "we may be spared the disturbance created by apparently new suggestions for radical departures from established lines of work" (22).

The conservatism of the project is quite apparent. Any departure from the canon established by Hartshorne is deemed "radical" and a "deviation" (102). "Geography is not an infant subject, born out of the womb of American geology a few decades ago, which each new generation of American students may change around at will" (29). Even more explicitly: "any efforts that require geography to change its essential character must be in vain; we cannot make over geography in any fundamental way, we can only fulfill that which it has been and is" (367). The future of geography is no less determined than its past, unless, of course, it is stolen by deviants and radicals.

Much as space is empty until filled, so too is time for Hartshorne. History is the means of filling time; it is the succession of discrete events occurring sequentially. Furthest from Hartshorne's purview is that history is a social process, that it results from struggle, that it is made—made, in fact, as geography is made. If geography is chorology, history is chronology. Thus Hartshorne's is a *continuist* history, a linear conceptual projection obviating any need for process, change, development, struggle or ultimately agency: *The Nature* "presented the evolution of geographic thought as on the whole in a continuous direction since the eighteenth century" (Hartshorne 1979, 73). Kant's foundational work (itself a continuation of the themes of antiquity) led naturally to Humboldt and Ritter and thence to Richthofen and Hettner. Hartshorne's servility to this history of his own making is best exemplified by his extraordinary claims for Kant:

> In the introduction to his lectures on physical geography, Immanuel Kant presented an outline of the division of scientific knowledge in which the position of geography is made logically clear. The point of view there developed has proved so satisfactory, to others as well as to this writer, both in leading to an understanding of the nature of geography and in *providing answers to all questions that have been raised,* that it seems worthwhile to quote at length from Kant's original statements (134, emphasis added.)

Until this decade, private histories still dominated in what passes for "the history of geographical thought," a pursuit that owes its perspective as well as its name to the conservative idealist tradition. As Stoddart (1986, 1) remarks, "under the influence of Kantian ideas, this methodology [history as chronology] was readily transferred to the history of geography as a whole." Hartshorne is the major figure in this process, and nowhere is his influence more alive and dominant than in the history of geographical thought. There are surely three objections to this hegemony of private history. In the first place, it perpetuates the defensive isolationism of the discipline as a whole through its internal focus and seemingly inexhaustible obsession with determining what is and is not geography. Second, such histories do a disservice to the discipline itself:

Inductivist and internalist history bears a heavy responsibility for reducing these activities to such simplistic terms in our standard histories that their intellectual content has been demeaned and our leading practitioners, while identified as the great men of the past, have been paradoxically reduced in stature (Stoddart 1986, 5).

Third, in the contemporary restructuring of the academic division of labor, private histories are afflicted with an additional layer of anachronism. The ideas and activities that built contemporary geography will best be kept alive not by careful cocooning but by active cross-fertilization and exposure to the transforming outside world. Ideas are like cultures; they are generally quite dead by the time they are committed to museums.

# Conclusion

Entrikin (1984, 393) has written of Sauer that he "was part of the antimodernist tradition in early twentieth-century American thought" which deplored the shift from an "agriculturally based rural society" to urban capitalism. The same assessment applies handsomely to Hartshorne, perhaps the consummate antimodernist. His is at root an eighteenth-century philosophy and a flatlander's geography; *The Nature* not only predates Einstein but, with the effort at a universal logic for geography, it predates Darwin and Marx as well. Explicitly absent in Hartshorne's vision are the realities of natural transformation and social change, evolution and revolution which even the conservative Mackinder could glimpse.

Certainly Hartshorne goes beyond Kant. His essential antinomy, highlighted by May (1972, 79), between geography as a systematic science and as a synthetic "queen of sciences," is a concession not only to neo-Kantian thinking but to the realities of the twentieth-century academic division of labor. It has been the conceit of most social sciences at one time or another that they, and they alone, were the grand synthesizers of the narrow toils of others, but it has been the particular conceit of geography to retreat to Kant for justification. Second, the contradiction in Hartshorne between a transcendental conceptual methodology and the appeal to a historical means for establishing that methodology encompasses a basic neo-Kantian tension, particularly prevalent in the Baden School.

But neither is Hartshorne a mechanical neo-Kantian, as is clear in one of the sharper ironies of his work. Like Kant, he seeks a unified methodology, abhors dualisms in geography, and is especially adept at blaming such dualisms upon "deviants" within the field or else external infiltrators (1958, 105). He seems to be nowhere cognizant of what the neo-Kantians knew only too well, namely that Kant himself was the unwitting source of dualism. Such a lacuna, it must be said, is at best naive. We can only assume that it results from Hartshorne's resolute lack of intellectual curiosity about the Kantian roots of his own thought.

The reassertion of dualism in Hartshorne, as in Kant, points up a more surprising aspect of Hartshorne's work. Hartshorne is no positivist, indeed the inspiration for his ideas, if it shares with positivism a common philosophical source, is equally pre-positivist. Yet especially in his hostility to history and ambivalent embrace of science, his conceptual formalism and the priority accorded to discrete geographical facts, Hartshorne glimpses an equally anti-historical positivism. Kant again is the crucial foundation. With positivism the unintended dualism of Subject and Object is now intended, and the philosophical priority is reversed; duality wins out over unity, objectivity over subjectivity. Within this framework, the positivist lexicon of things and facts, causes and empirical observations (if now deemed to exist in objective reality), can be traced back to Kant.

Yet this may not be such a surprising result after all. As Russell (1945, 718) points out, "Kant's inconsistencies were such as to make it inevitable that philosophers who were influenced by him should develop rapidly either in the empirical or in the absolutist direction." Subject and Object are either wrenched apart, as with August Comte and the ensuing positivist tradition, or else rendered indistinguishable, as with Hegel. The neo-Kantian revival was in large part aimed against a Hegelian hegemony in mid-century German philosophy. It is worth recalling R. G. Collingwood's (1957, 169) jibe at Rickert who, while critiquing positivism, also adopted some if its central assumptions: "Rickert regards nature . . . as cut up into separate facts and he goes on to deform history by regarding it in a similar way as an assemblage of individual facts. . . ." As regards Hettner, Hartshorne (138 fn.) himself reiterates the characterization of his work as "based on the 'positivistic liberalism' of the nineteenth century."

Given the diversity of early twentieth-century neo-Kantianism, there was surely no inevitability to the path Hartshorne took. Durkheim, Park and Max Weber, all in very different ways, went on to become major inaugural figures in modern sociology; the early Frankfurt School sought a rapprochement between Hegel and Marx that leaned heavily on a critical reading of Kant (Jay 1973; Schmidt 1971). Weber in particular is an interesting contrast to Hartshorne since he deliberately took his ground on Rickert's methodology and tried to apply it to sociology (Oakes 1987; Willey 1978). He too confronted Rickert's dilemma of the historical individual and the necessity for a theory of value, but where Hartshorne fastened on the farm as the appropriate geographical individual, Weber derived the notion of the "ideal type."

The point here is not that Rickert and Weber got the right answer while Hartshorne somehow erred; neither Rickert nor Weber resolved the question satisfactorily, which perhaps explains its reappearance today at the core of Habermas's (1976, 1979) theory of communicative competence, in Giddens's (1981, 1984) structuration theory,[8] and in Lyotard's (1984) appeal to a theory of justice. Rather, their answers opened up new avenues of research, while Hartshorne's resort to the farm was met with a silence that echoed the irrelevance of his solution. Such an agenda closed off the majority of interesting geographical questions associated with twentieth-century urban and industrial expansion or the restructuring of the global political economy under the Pax Americana. But, of course, it was regions, not farms that really interested Hartshorne; regions were the functioning if not the epistemological "geographical individuals" in his geography, but insofar as no systematic classification of region could be assumed and insofar as his conception of region was idealist and anti-historical and retained the agricultural bias of the "farm," the regional concept was equally a cul-de-sac. Only now is a more materialist basis for regional geography beginning to be reconstructed (Thrift 1983, 38–42; Pudup 1988).

Other forces contributed to the conservative isolationism that *The Nature* encouraged in geography. Although Hartshorne had certainly carried out substantive geographical research, especially in agricultural and political topics, *The Nature* did not easily connect with this work (nor indeed with the regional geographies of other practicing geographers), thereby codifying a more abrupt disjunction between methodology and research than resulted elsewhere. Hartshorne himself seems not to have followed up the results of his own argument with substantive research on regions as complexes of farms, and indeed it is difficult to point to any major reorientation of regional geography as a result of *The Nature*. This is precisely the point. Its lack of transformative effect should not be taken as a sign of the work's ephemerality. In sanctifying the status quo in the research sphere, it was a spectacular success, while its difficulty simultaneously put methodological discussion beyond the ability and interest of most practicing geographers. The book was revered as a talisman of geography's arrival.

The irony is that this retreat to methodology took place on the basis of unexamined philosophical assumptions. How could Weber have followed Rickert more closely (which he did) and yet have avoided the regressive consequences that came to geography via Hartshorne? The answer is implied in the question. Weber's agreement with Rickert, the positioning of his social theory vis-à-vis Kant, and his search for a normative theory of value, were, regardless of their destination, the result of Weber's own intensive inquiry. By no means indisputable, this philosophical security nevertheless released him to migrate freely between contemporary political and social questions on the one side and philosophical issues on the other. Hartshorne, by contrast, achieves no such freedom. He cannot garner even partial reconciliation with the Kantian philosophical tradition as long as he shrugs at Kant and neo-Kantianism, brackets them off as of no concern, beyond geography, philosophy. He neither gets Kant out of his system nor learns to live with him, precisely because he denies that the philosophical Kant was ever there. Hartshorne's true role in the history of geography was to take us back around the philosophical wheel to the eighteenth century.

In fairness to Hartshorne, it must be said that he did not accomplish the isolation of mid-twentieth century American geography single-handedly. The great retreat was already underway when he received his Ph.D. from Chicago in 1924. That he was not effectively challenged until the 1950s was also a broader disciplinary responsibility. If Hartshorne attempted to narrow the foundation of the discipline, it is also worth remembering that he inherited a very narrow philosophical tradition, in the U.S. if not in continental Europe. That Sauer may reasonably be considered his major intellectual adversary of the time is a sobering thought, for Sauer shared much with Hartshorne—both were neo-Kantians whose main inspiration came from German geography, Hettner specifically. If Sauer was more historically, culturally and materially oriented, these differences were not dramatic in the larger intellectual scope of the period.

The situation today is dramatically different. Not only are many geographers more cognizant of the myriad shifts, evolutions and new directions in the intellectual mainstream, especially in social theory, but the discipline is internally more diverse. So why is Hartshorne's work now drawing attention as part of a broader revision and reassessment of the history of geography?

In the first place, the humanist critique of positivism recentered the human individual as a coherent geographical subject; meaning, uniqueness and individuality are central to this work and the Kantian and neo-Kantian traditions are being explored for insights (Livingstone and Harrison 1981; Entrikin 1985). Second, there is the reaction to marxism. In retrospect, the ability of marxian theory to define so much of the research frontier in the 1970s and early 1980s is surely surprising given the increasing political conservatism of the period. The paucity of competing social theories, even after the emergence of positivism, meant that in geography, to a disproportionate extent, marxian research and social theory were for a short time virtually synonymous. The 1980s, however, have seen a steadfast deconstruction of marxian theory throughout the social sciences; human geography, it seems, has put itself through nearly a century of social thought in less than three decades. Here, as throughout the social sciences, we can now observe a highly eclectic and politically defused body of social theory. Within geography, there has been a certain convergence incorporating some of the concepts initially adopted by humanists, but with the demand for a more searching theoretical exposition. Hence Giddens's efforts to recenter individual agency against structuralism. Previously attracted to Althusser's structuralism, Massey's (1985, 19) suggestion that "the unique is back on the agenda" resonates strongly among geographers of highly diverse perspectives. The reawakening of interest in Hartshorne has to be understood in this context.

Matching the current social, political and geographical restructuring is a cultural and intellectual restructuring of modernism, sufficiently profound that notions of post-modernism have become central to social theory. It is not only the structure and politics of marxism that is being deconstructed; the web of disciplinary boundaries defining the century-old academic division of labor is itself showing signs of melting into air. Michael Eliot-Hurst's (oft-misdenoted) *de*-definition of geography (1980) is being accomplished before our eyes, less through the radical agency of any disciplinary coup-d'état and more through an internal restructuring throughout the social sciences. With it comes a recentering of geographical space in critical social theory (Soja 1989).

The parallels with the *fin de siècle* and the emergence of high modernism ought not to be overdrawn, but now as then an intellectual restructuring is matched by a material restructuring of the conditions of society (Harvey 1989).

A Kantian revival is intertwined with a rediscovery of space. The earlier Kantian revival of a century ago also embodied specific political aspirations including the liberal vision of an amalgamated Germany. The expansion and institution-alization of geography proceeded apace after Bismarck's 1871 confirmation as first chancellor of the German Empire. In France, neo-Kantianism "was utilized to reject competing ideologies and to promote the secularized, individualistic, and nationalistic ideology of the newly established Third Republic" (Berdoulay 1978, 80).

The form and trajectory of contemporary social theory resembles a tricky gestalt. From one angle, the eclectic diversity and emphasis on particularity are intended as an antidote to "totalizing" discourses that are held to stress unity and generality over difference and uniqueness. These new departures exhort the political empowerment of particular individuals and agents as against social structure and revolutionary organization. From a different angle, the resurgence of neo-Kantian themes is implicated in two specific trends: first, the privileging of individual and local over global or societal empowerment; second, the so-called linguistic turn and a disproportionate emphasis on the "proactive" ability of individuals to construct their own worlds through the construction of discourse and concept formation. Together these trends threaten a new set of dualisms with equally political consequences. They accentuate the fragmenta-tion and depoliticization of people united on the basis of class, gender, race and so forth. Epistemological difference constructs social distinction. The current reorientation toward Kant and neo-Kantian ideas is in part the political means for getting the radical genie back in the bottle, not just in geography but throughout the social sciences. The rehabilitation of Hartshorne in this context, however partial and circumspect, would indulge a new anti-modernism which, while it may never attain the power to return geography to the museum, might contribute to a deepening of "the great desert of the American mind" (Kirby 1989) and the larger isolation of social science from the world it would under-stand.

## Acknowledgment

In addition to several anonymous referees, I would like to thank Nick Entrikin, David Harvey, Don Mitchell, Julie Tuason, and John E. Brush for very useful comments. John Holmes of the University of Queensland also offered comments as well as a more contemplative environment for writing. To Professor Hart-shorne himself, I am grateful for spirited criticisms on an earlier draft.

## Notes

1. Among more contextual histories of geography, many are essentially intellectual histories and critiques. See for example Stoddart (1986); Berdoulay (1978, 1981); Glick (1983); Entrikin (1984); Livingstone (1987); Solot (1986). Others have adopted a more materialist concern to link the intellectual history of the discipline to broader questions

of social and geographical change: Hudson (1977); Capel (1981); Smith (1984a, 1988); Peet (1985); Stoddart (1986); Breitbart (1981); Powell (1988); Godlewska (1989); Kirby (1989 forthcoming).

2. Isaiah Bowman to James Truslow Adams, 2 August 1924, Isaiah Bowman Collection, Johns Hopkins University.

3. Citations in this paper that refer only to page numbers are to Hartshorne, 1961 ed.

4. The most explicit evocation of this geopolitical metaphor for knowledge comes from the French philosopher Louis Althusser who also evinces his own version of the *hiatus irrationalis* with an unbridgeable distinction between the "real object" and the "object of knowledge." The extent to which Althusser smuggles a Kantian contraband into his reconstructed structuralism has so far gone unremarked by Althusser's critics, many of whom also draw on Kant as inspiration for a more humanist vision (Althusser and Balibar 1970, 67; Althusser 1971, 14–17 passim; for a critique see Smith 1980).

5. Commenting on this metaphysical argument, Russell (1945, 716) writes: "Its premiss is 'space is imagined . . . as an infinite *given* magnitude.' This is the view of a person living in a flat country, like that of Königsberg; I do not see how an inhabitant of an Alpine valley could adopt it."

6. In the abstract to *The Nature* added in 1946, he softens the impact by adding "a farmer's field or a city block, . . . a factory or even a city" (xii) as plausible "primary Wholes." For *Perspective,* he reconstructs the discussion of regions, avoids any discussion of farms or other "primary Wholes," and in general backs away from dealing with the question. He recognizes the "vagueness" of the term, region, and admits there may be no formal definition (1959, 108–45).

7. In *Perspective,* Hartshorne cautiously incorporates a more liberal discussion of relationships, in part easing the contradiction between his conception of relative location and disavowal of relationships as an epistemological basis for geography (1959, 113–42).

8. It is worth noting too that Gidden's (1981, 1984) efforts to recenter geographical space in social theory resort to a Kantian ontological assertion of the equivalence of space and time that remains somewhat disconnected from the practical exposition of structuration theory.

# References

Althusser, Louis. 1971. *Lenin and philosophy.* New York: Monthly Review Press.

————, and Balibar, Etiene. 1970. *Reading capital.* London: New Left Books.

Banse, Ewald. 1928. *Landschaft und Seele. Neue Wege der Untersuchung und Gestaltung.* Munich: R. Oldenbourg.

Barrows, Harlan. 1923. Geography as human ecology. *Annals of the Association of American Geographers* 13:1–14.

Beer, M. 1898. Der moderne Englische Imperialismus. *Die Neue Zeit* 41:300–306.

Berdoulay, Vincent. 1978. The Vidal-Durkheim debate. In *Humanistic geography: Prospects and problems,* ed. David Ley and Marwyn S. Samuels, pp. 77–90. Chicago: Maaroufa Press.

————. 1981. The contextual approach. In *Geography, ideology and social concern,* ed. David Stoddart, pp. 8–16. Oxford: Basil Blackwell.

Berman, Marshall. 1982. *All that is solid melts into air: The experience of modernity.* New York: Simon and Schuster.

Bowman, Isaiah. 1934. *Geography in relation to the social sciences.* New York: Charles Scribner and Son.

Breitbart, Myrna. 1981. Peter Kropotkin, the Anarchist geographer. In *Geography, ideology and social concern,* ed. David Stoddart, pp. 134–64. Oxford: Basil Blackwell.

Capel, Horacio. 1981. Institutionalization of geography and strategies of change. In

*Geography, ideology and social concern*, ed. David Stoddart, pp. 37–69. Oxford: Basil Blackwell.

Cassirer, Ernst. 1981. *Kant's life and thought*. New Haven, CT: Yale University Press.

Collingwood, R. G. 1957. *The idea of history*. New York: Oxford Galaxy Books.

Cosgrove, Denis. 1984. *Social formation and symbolic landscape*. London: Croom Helm.

———. 1985. Prospect, perspective and the evolution of the landscape idea. *Transactions, Institute of British Geographers* N.S. 10:45–62.

———, and Daniels, Stephen J., eds. 1988. *Iconography of landscape*. Cambridge: Cambridge University Press.

Cronon, William. 1983. *Changes in the land. Indians, colonists, and the ecology of New England*. New York: Hill & Wang.

Dodge, Stanley. 1936. The chorology of the Claremont-Springfield region in the Upper Connecticut Valley in New Hampshire and Vermont. *Papers of the Michigan Academy of Science, Arts and Letters* 22:335–53.

Eliot-Hurst, Michael. 1980. Geography, social science and society: Towards a de-definition. *Australian Geographical Studies* 18:3–21.

Entrikin, J. Nicholas. 1980. Robert Park's human ecology and human geography. *Annals of the Association of American Geographers* 70:43–58.

———. 1984. Carl O. Sauer, philosopher in spite of himself. *Geographical Review* 74: 387–408.

———. 1985. Humanism, naturalism and geographical thought. *Geographical Analysis* 17:243–47.

Garnett, Christopher Browne. 1939. *The Kantian philosophy of space*. Port Washington, NY: Kennikat Press.

Gerland, Georg. 1905. Immanuel Kant, seine geographischen und anthropologischen Arbeiten. *Kant-Studien* 19:1–43, 417–547.

Giddens, Anthony. 1981. *A contemporary critique of historical materialism*. Berkeley: University of California Press.

———. 1984. *The constitution of society*. Berkeley: University of California Press.

Glick, Thomas F. 1983. In search of geography. *Isis* 74(271):92–97.

Godlewska, Anne. 1989. Traditions, crisis and new paradigms in the rise of the modern French discipline of geography 1760–1850. *Annals of the Association of American Geographers* 79:192–213.

Habermas, Jürgen. 1975. *Legitimation crisis*. Boston: Beacon Press.

———. 1979. *Communication and the evolution of society*. Boston: Beacon Press.

Hart, John Fraser. 1982. The highest form of the geographer's art. *Annals of the Association of American Geographers* 72:1–29.

Hartshorne, Richard. 1955. "Exceptionalism in geography" re-examined. *Annals of the Association of American Geographers* 45:205–44.

———. 1958. The concept of geography as a science of space, from Kant and Humboldt to Hettner. *Annals of the Association of American Geographers* 48:97–108.

———. 1961 ed. *The Nature of Geography. A critical survey of current thought in the light of the past*. Lancaster, PA: Association of American Geographers.

———. 1972. Review. "Kant's concept of geography" by J. A. May. *Canadian Geographer* 16:77–79.

———. 1979. Notes toward a bibliobiography of *The Nature of Geography. Annals of the Association of American Geographers* 69:63–76.

———. 1988. Hettner's exceptionalism—fact or fiction. *History of Geography Journal* 6: 1–4.

Harvey, David. 1969. *Explanation in geography*. London: Edward Arnold.

———. 1989 forthcoming. *The condition of postmodernity*. Oxford: Basil Blackwell.

Hettner, Alfred. 1927. *Die Geographie, ihre Geschichte, ihr Wesen, und ihre Methoden*. Breslau: Hirt.

Hudson, Brian. 1977. The new geography and the new imperialism, 1870–1918. *Antipode* 9(2):12–19.

Jay, Martin. 1973. *The dialectical imagination. A history of the Frankfurt School and the Institute of Social Research 1923–1950*. London: Heinemann.

Kant, Immanuel. 1919 ed. *Critique of pure reason*. London: Macmillan.

———. 1923 ed. Physische geographie. *Gesammelte Schriften* Bd. 9:151–436.

Kern, Stephen. 1983. *The culture of time and space*. London: Wiedenfield and Nicholson.

Kirby, Andrew. 1989 forthcoming. The great desert of the American mind. Concepts of space and time and their historiographic implications. In *The estate of social knowledge*, ed. JoAnne Brown and David van Keuren. Baltimore: Johns Hopkins University Press.

Leighly, John. 1937. Some comments on contemporary geographic methods. *Annals of the Association of American Geographers* 27:125–41.

Lenin, Vladimir. 1917. *Imperialism, the highest stage of capitalism*. Beijing: Progress Publishers.

Livingstone, David N. 1987. *Nathanial Southgate Shaler and the culture of American science*. Tuscaloosa, AL: University of Alabama Press.

———, and Harrison, R. T. 1981. Immanuel Kant, subjectivism, and human geography: A preliminary investigation. *Transactions, Institute of British Geographers* N.S. 6:359–74.

Lyotard, Jean-Francois. 1984. *The postmodern condition: A report on knowledge*. Minneapolis: University of Minnesota Press.

Mackinder, H. J. 1904. The geographical pivot of history. *Geographical Journal* 23:421–37.

Marshall, Alfred. 1890. *Principles of economics*. London: Macmillan.

Marx, Karl, and Engels, Friedrich. 1955 ed. *The communist manifesto*. New York: Meredith Corporation.

Massey, Doreen. 1985. New directions in space. In *Social relations and spatial structures*, ed. Derek Gregory and John Urry, pp. 265–95. Basingstoke, England: Macmillan.

May, J. A. 1970. *Kant's concept of geography and its relation to recent geographical thought*. Toronto: University of Toronto Press.

———. 1972. A reply to Professor Hartshorne. *Canadian Geographer* 16:79–81.

Oakes, Guy. 1987. Weber and the Southwest German School. The genesis of the concept of the historical individual. In *Max Weber and his contemporaries*, ed. Wolfgang J. Mommsen and Jürgen Osterhammel, pp. 434–46. London: George Allen and Unwin.

Olwig, Kenneth. 1984. *Nature's ideological landscape*. London: George Allen and Unwin.

Park, Robert. 1925. *The city*. Chicago: University of Chicago Press.

Peet, Richard. 1985. The social origins of environmental determinism. *Annals of the Association of American Geographers* 75:309–33.

Powell, Joe. 1988. *An historical geography of modern Australia*. Cambridge: Cambridge University Press.

Prunty, Merle C. 1979. Southern geography. *Annals of the Association of American Geographers* 69:53–58.

Pudup, Mary Beth. 1988. Arguments within regional geography. *Progress in Human Geography* 12:369–90.

Rickert, Heinrich. 1962. *Science and history. A critique of positivist epistemology*. Princeton, NJ: Van Nostrand.

Rorty, Richard. 1979. *Philosophy and the mirror of nature*. Princeton, NJ: Princeton University Press.

Russell, Bertrand. 1945. *History of western philosophy*. New York: Simon and Schuster.

Sauer, Carl. 1941. Foreword to historical geography. *Annals of the Association of American Geographers* 31:1–24.

Sayer, R. A. 1984. *Method in social science: A realist approach*. London: Hutchinson.

Schaefer, Fred. 1953. Exceptionalism in geography: A methodological examination. *Annals of the Association of American Geographers* 43:226–49.

**Schmidt, Alfred.** 1971. *The concept of nature in Marx.* London: New Left Books.

**Schorske, Carl.** 1980. *Fin de siècle Vienna: Politics and culture.* London: Weidenfeld and Nicolson.

**Smith, Neil.** 1980. Symptomatic silence in Althusser: The concept of nature and the unity of science. *Science and Society* 44:58–81.

————. 1984a. Political geographers of the past. Isaiah Bowman: Political geography and geopolitics. *Political Geography Quarterly* 3:69–76.

————. 1984b. *Uneven development: Nature, capital and the production of space.* Oxford: Basil Blackwell.

————. 1988. For a history of geography. *Annals of the Association of American Geographers* 78:159–63.

**Soja, Edward W.** 1989. *Postmodern geographies: The reassertion of space in critical social theory.* London: Verso.

**Solot, Michael.** 1986. Carl Sauer and cultural evolution. *Annals of the Association of American Geographers* 76:508–20.

**Stoddart, David R.** 1966. Darwin's impact on geography. *Annals of the Association of American Geographers* 56:683–98.

————. 1986. *On geography.* Oxford: Basil Blackwell.

**Supan, Alexander.** 1906. *Die territoriale Entwicklung der Europächen Kolonien.* Berlin, Gotha: J. Perthes.

**Thrift, Nigel.** 1983. On the determination of social action in space and time. *Environment and Planning: Society and Space* 1:23–57.

**Willey, Thomas E.** 1978. *Back to Kant: The revival of Kantianism in German social and historical thought, 1860–1914.* Detroit: Wayne State University Press.

**Windelband, Wilhelm.** 1980. Rectoral address (Strasbourg 1894). *History and Theory* 19: 169–85.

# Sameness and Difference: Hartshorne's *The Nature of Geography* and Geography as Areal Variation

JOHN A. AGNEW

Department of Geography, Syracuse University, Syracuse, NY 13244

Reinterpreting the writings of Richard Hartshorne, in particular his *The Nature of Geography*, has been a common feature of American geography over the past forty years. Schaefer (1953) and Harvey (1969) saw Hartshorne as a neo-Kantian idealist because of his presumed emphasis solely on the "uniqueness" of locations. Sack (1974), Guelke (1978), and Gregory (1978) have, in very different ways, defined Hartshorne as a positivist (or proto-positivist) because of an alleged "implicit" acceptance of a positivist model of science. This paper provides a rereading of Hartshorne's *The Nature* which suggests that it can be seen as presenting a very different view of geography than is usually credited to it. Stated boldly, the thesis is that elements of a view of geography as areal *variation* rather than areal *differentiation* can be found in *The Nature*. The key claim is that Hartshorne rejected the simple polarity between sameness and difference or the generic and the specific that has bedeviled so much methodological discussion in geography, including most interpretations of *The Nature*. Hartshorne was, unfortunately, unable to effectively demonstrate how this rejection could work in practice. As a consequence, in his proposal of a regional focus for geography, he has been seen as either defending an idiographic-singular conception of regions for the practice of regional geography or not having much to offer to any "new" regional geography.

From one point of view, this paper could be seen as an attempt at appealing to a "classic" in order to justify recent attempts at rehabilitating regional geography. Hartshorne himself used German "authorities" (in particular, Hettner) to justify his arguments about the nature of geography. Although this paper does purport to find a heretofore "hidden" view of geography in *The Nature*, it does so only as a point of departure for a more systematic argument that cannot itself be found in any of Hartshorne's writings.

The first section of the paper provides a critical evaluation of Hartshorne's discussion of "what kind of a science is geography?" in terms of the sameness-difference polarity. A second section proceeds from Hartshorne's discussion by proposing a realist theory of place in which the polarity is transcended. To illustrate the continuity and discontinuity between this perspective and that of *The Nature,* a third section uses recent research on the geography of Italian electoral politics to demonstrate how geographers can combine an interest in sameness and difference rather than emphasizing only one or the other.

# Sameness and Difference

## The Nature of Geography

Hartshorne's central claim about geography is its integrative or synthetic purpose:

> Geography does not claim any particular phenomena as distinctly its own, but rather studies all phenomena that are significantly integrated in the areas which it studies, regardless of the fact that those phenomena may be of concern to other students from a different point of view (1939, 372).

This view sometimes leads Hartshorne to a particular corollary:

> the interest of the geographer is not in the phenomena themselves, their origins and processes, but in the relations which they have to other geographic features (i.e., features significant in areal differentiation) (1939, 425–26).

Much of Hartshorne's argument in *The Nature* does not involve this excessive focus on areal differentiation. Indeed, his presentation builds a case for geography as the study of areal variation, sameness *and* difference, between areas rather than simply areal differentiation. Hartshorne's problem was that, without an interest in "the phenomena themselves," what there was to be analyzed geographically, he had no way of resolving the sameness-difference dualism by identifying the *causes* of a particular pattern of areal variation. Hence came the ease with which subsequent interpreters could regard *The Nature* as licensing an exclusive emphasis on areal differentiation and regions as containers of an endless variety of phenomena.

In this section, an interpretation of *The Nature* in terms of areal variation is offered. Then some reasons are suggested why this interpretation has not been offered before by others. First of all, Hartshorne offers an argument for the importance of generic principles and abstract categories in geography. On p. 383, he writes:

> a geography which was content with studying only the individual characteristics of its phenomena and their relationships and did not utilize every opportunity to develop generic concepts and universal principles would be failing in one of the main standards of science.

But he adds later (387) (and this sets him apart from the conventional positivist/logical empiricist rendering of science characteristic of his era):

Generic concepts are not an end in themselves but merely a scientific tool whose particular purpose varies according to the point of view of the different branches of science.

These two quotations illustrate a common tendency in *The Nature* to try to provide a set of minimum criteria for geography as a "scientific" enterprise: (1) a commitment to empirical observation, (2) objectivity, (3) a concern for universality or generic principles, and (4) a concern for the systematic ordering of existing knowledge (Entrikin 1981, 4). These cannot be reduced to those criteria provided by positivist philosophers of science. Indeed, throughout *The Nature*, Hartshorne refers to the *practices* of other sciences rather than to the pronouncements of positivist philosophers. Geography, in integrating facts about areas with classificatory concepts, provides a type of knowledge that is similar to that produced by other sciences.

Second, in distinguishing the "nomothetic" or law-making sciences from the "idiographic" sciences, those concerned with the *einmalige*, the unique, Hartshorne does not define geography as completely one or the other. Although he refers to Windelband and Rickert as the originators of this division, he does not endorse it. Rather, he argues that the "two aspects of scientific knowledge are present in all branches of science" (379). He continues:

The common idea that generalizations and laws themselves are the purpose of science is characterized by Hettner as an extraordinary adherence to medieval scholasticism. On the contrary, they are [to Hettner] "merely the means to the ultimate purpose, which is the knowledge of actual reality, the individual facts, either conditions or events" (379).

Hartshorne illustrates this point using the example of astronomy:

The astronomer develops laws of celestial mechanics not in order to prove that the universe is governed by law—which is a philosophical rather than a scientific thesis—but in order that he can rightly understand the motions of the heavenly bodies. He does not forget his interest in the latter as individuals (1939, 379).

Here again Hartshorne is rejecting thinking in terms of conventional polarities. Regions are "individuals" but they are constituted by the causal connections between the phenomena that define them. This view hardly demonstrates the wholehearted commitment to the "idiographic" claimed by some interpreters!

Third, Hartshorne is expressly critical of an "enthusiasm for patterns as such" which he claims was characteristic of the landscape studies of the Berkeley School (235):

By reducing the element of relative location to a purely secondary position it [the landscape approach] tends to neglect the very essence of geographic thought— integration of phenomena in spatial associations. In consequence there develops an uncritical enthusiasm for patterns as such, regardless of their significance, and an over-emphasis on form in contrast to function which tends to slight those characteristics of areas whose importance is not represented proportionately by material objects.

Rather than regarding regions solely in terms of their *einmalige* combinations of interrelated phenomena, Hartshorne proposes that:

the significance of patterns depends entirely on the extent to which they depict significant relations in the location of different places in relation to each other (1939, 226).

This would again appear to reveal an interest on Hartshorne's part in the nature and causes of place-to-place similarity as much as in regional specificity.

Fourth, in his discussion of the uniqueness of regions Hartshorne returns to the role of generic concepts. The focus on the unique combinations of phenomena that give a region its specific character:

does not mean that we cannot utilize generic concepts that express marked similarities in the characteristics of different areas. On the contrary, it is of great value to discover element-complexes and combinations of several element-complexes repeated in different areas (1939, 396).

Here Hartshorne is at his most explicit in identifying regions with element-complexes or combinations of phenomena that are not unique but also occur in other areas.

Fifth, Hartshorne emphasizes scale as the essential context for understanding sameness and difference, although he does not use those words:

The localities of a larger region do not represent independent specimens of a species but simply similar parts of a whole, whose similarities are based in major part on the fact that they are but parts of a whole (1939, 395).

In other words, localities and larger regions cannot be understood in isolation from one another but only in terms of their similarities and differences.

Sixth, and finally, sameness and difference apply to the combinations of phenomena, not to areas in themselves.

In other words, the attempt to develop generic concepts *about* areas—and, on that basis, to compare areas in themselves and develop principles of their relations—rests on the fallacious assumption of the area as an object or phenomenon (1939, 395).

Hartshorne wanted to avoid "areal fetishism," separating area from phenomena and treating areas as things in themselves, as he built his case for regions as unique combinations of phenomena that can only be understood as such in relation to one another and to larger areas. Consequently, his logic suggests that in order to recognize geographical difference one must necessarily examine geographical similarity.

## Barriers to this Interpretation

In a number of ways, if very unsystematically, Hartshorne presented, in *The Nature*, parts of an argument for geography as areal variation rather than areal differentiation. If this interpretation has merit, why has it not been suggested before? A number of reasons are possible though none can be regarded as definitive.

In the first place, Hartshorne's repeated use of the term areal differentiation has, not surprisingly, deflected commentators away from the complexity of his presentation towards a search for quotations that are consistent with the emblematic term. The fact that *The Nature* became a weapon in battles to have geography accepted as a separate university "subject" rather than a centerpiece for intellectual debate perhaps encouraged this simplification (Meinig 1988). Thus, to Johnston (1983, 43) "Hartshorne argued forcefully that the focus of geography is areal differentiation, the mosaic of separate landscapes on the earth's surface." Johnston then offers a series of quotations that provide evidence for this view. Interestingly, Hartshorne commented, in a footnote added to the seventh imprint of *The Nature* (1960), that:

> The term "areal differentiation" as used here and in innumerable other places in this volume, has been found to lead easily to misunderstandings, so that the author now uses rather the term "areal variation" (footnote 35, 237).

Hartshorne then directs the reader to his *Perspective on The Nature of Geography* (1959, 12–21).

It is probably also the case that associating Hartshorne with areal differentiation as indistinguishable from an idiographic viewpoint was also part of the labeling and oversimplification of methodological positions that went on in American geography in the 1960s and 1970s (Entrikin 1985). A new generation that wanted nothing to do with the largely descriptive accounts of conventional regional geography could discard the idea of regional geography more easily if it could be characterized as intrinsically concerned with uniqueness and uninterested in generic principles. As recently as 1979, some of the "young turks" of the 1960s were still arguing this way (Gould 1979).

A third reason stems from Hartshorne's failure to define explicitly what he means by science. Indeed, he openly refused to do so (396), thus allowing his critics to impose a positivist/logical empiricist one. This void has left areal variation without an "overt scientific" defense, but one is in fact possible. Causal explanation, as Hartshorne emphasized (Entrikin 1981, 6–7), need not be restricted solely to relationships between *classes* of objects or events, as it is in a logical empiricist account. In a scientific realist account, for example, causal explanation involves positing causal or generative mechanisms that produce observable outcomes including individual or unique combinations of phenomena (Outhwaite 1987). From this point of view, theory is no longer associated with generality in the sense of repeated series of events everywhere. Rather, as Sayer (1988, 6) puts it: "Generality in the sense of extent of occurrence . . . depends upon how common instances of the object are, and upon the circumstances or conditions in which objects exist, these determining whether the causal powers and liabilities of objects are activated." There is thus no need to see geography as construed by Hartshorne as unscientific or anti-causal per se.

Above all, Hartshorne's frequent reference to *significant* phenomena and relations as the defining attributes of specific regions is no substitute for attention to theory that would select and order these attributes. Hartshorne's lack of interest in "the phenomena themselves, their origins and processes" (425) prevented the possibility of his actually defining significance and thus stating what separates a region from others, both in the present and the past. To resolve the sameness-difference dualism requires both knowledge of the significant spatially proximate and causally linked phenomena upon which any particular regional study is based and, especially, a way of explaining their areal variation.

Hartshorne's "problem" was his view of geography as a field that dealt with *all* the characteristics of areas (physical, social, economic, political, etc.) in combination rather than as a subject that analyzed the processes which produce combinations of phenomena in areas. Significance is hard to define when a field is seen as a factual vacuum cleaner, searching for a "special," totalistic or integralistic knowledge, rather than a subject involved with specific problems which are shared with and can draw from the literatures of other fields (Gambi 1973, 205). Hartshorne's "integralism" was thus a major barrier to his ever answering satisfactorily the question of significance and resolving the sameness-difference polarity in a theoretically coherent manner.

# Causes and Outcomes in Areal Variation

In this section, an approach to resolving "Hartshorne's problem" responds explicitly to the lacunae in Hartshorne's case for areal variation. In a realist view, science is about the causal powers and liabilities of objects and people rather than relationships between discrete events (Harre and Madden 1975). A causal claim is, in this view, what an object or person is like and what it *can* do. Causal powers and liabilities can be attributed independently of any specific pattern of events. As Sayer (1984, 99) argues: "although causal powers exist necessarily by virtue of the nature of the objects [or people] which possess them, it is contingent whether they are ever activated or exercised." This means that, with respect to people, the operation of causal power can produce different outcomes because of the mediating effects of human agency. In other words, the spatially differential activation of causal powers produces areal variation. Rather than generalizing from universal propositions about process to a universal geographical form or outcome, a focus on areal variation can provide a frame of reference for examining the relationship between causes and outcomes *without* a presumption of universality in outcomes.

## Geo-historical Synthesis

This conception of the relationship between causes and outcomes has been used in some recent writing that attempts to create a revived regional geography or, perhaps more accurately given the emphasis on process, a new "geo-historical synthesis" (Sayer 1988, 2). Much of this literature uses the terms "place" and "locality" rather than region. The meso-scale region is not privileged as the only "level" of geographical sameness and difference in outcomes, as the example below on Italian electoral geography demonstrates. The new work is much more oriented to the phenomena of the social sciences, especially sociology, than was the "old" regional geography (for a critical review, see Jonas 1988). It is oriented to areal variation in specific economic, social, and political phenomena rather than the "total" regional accounts Hartshorne had in mind (see e.g., Massey 1984; Pred 1984; Urry 1987; Warf 1988; Warde 1988). There is some continuity with Hartshorne, however. In addition to the focus on areal variation, there is agreement on the definition of geography as a field focusing on how phenomena relate in space rather than a field with its own "phenomenon": space as such.

One approach within this genre, concerned with the geography of political activity, develops the realist perspective in a specific way by using a concept of place that has three major elements: locale, the settings for social interaction in which relatively enduring social relations are constituted; location, the effects of economic and social processes operating at wider scales upon locales; and sense of place, the identification with a place that can follow from living in it (Agnew 1987). By way of example, home, work, school, church and so on form nodes around which everyday life circulates, economic interests are defined and political capacities arise. Except for the most isolated places, however, the "time-space distanciation" of social interaction always involves the expansion of social relations beyond the immediately local. Place is *not* the same as locality. The most important factor in the historical expansion of "external" ties has been the growth in the social division of labor and the emergence of a class society (Giddens 1979). For the working class, for example, heavily dependent on a limited range of local job opportunities, interaction is still locally dependent, even with the growth of phenomena such as commuting and leisure travel. Managers and business owners, particularly those involved with firms that have local markets, likewise face a spatially constrained environment because of their local dependence (Cox and Mair 1988). The *relatively enduring* social relations upon which the identities and interests of most people are based remain grounded in an everyday life of narrowly circumscribed routines, schedules, capacities and interests. Yet the increasingly global organization of production, the increased homogenization of human practices through the influence of mass media and national education systems, and the "surveillance" of national governments over their populations have helped to make the practices reproduced in different places more and more alike.

These tendencies toward homogenization do not mean that place as a particular locational intersection of important settings for social action has lost coherence or meaning. Common experiences engendered by the social forces of nationalization and globalization are still mediated by local ones. Most people still follow well-worn local paths in their daily existence. Though issues with a wider frame of geographical reference than the strictly local have increased in number and significance, they take on meaning as they relate to the priorities and orientations that emanate from the practical reasoning of everyday life.

## Place and Politics

With respect to the geography of political activity, a topic taken up later for a particular national context, some specific causal mechanisms seem especially important in general terms but produce different outcomes in different places because of the active constitutive effects of distinctive locales. A number of these can be identified in contemporary Europe and North America. Perhaps most significantly, the social division of labor takes a spatially-differentiated form that evolves in rhythm with changes in the world economy (Taylor 1988). Whether decisions are made locally or between places, there is an unevenness in the spatial distribution of investment, skills, input sources and markets (Pred 1984, 284). Different places have different relationships to the world economy: some operate as the "production outposts" of multinational firms; some are more insulated from the world economy with an orientation towards regional or national markets threatened by "foreign" competition; others are themselves the headquarters for "international empires" (Massey 1984, 299). These relationships are of fundamental importance with respect to the organization of social life and the definition of dominant local interests.

Second, changing technologies and shifting forms of business organization produce different "spatial requirements" for many firms. In the contemporary U.S., for example, intra-firm site and inter-firm agglomeration economies (returns to geographical concentration) have declined across a number of industrial sectors in the face of new technological linkages between firm headquarters and branch plants (Gertler 1986). At the same time, many firms have been faced by an increasingly competitive world economy. While opening up certain new employment in some places, these processes have removed other kinds of employment elsewhere, partly depending upon local conditions such as "business climates" and local labor histories (Agnew 1988a). As a result, seemingly permanent geographical patterns of political affiliation and activity have been disrupted in diverse ways.

Third, all places are encapsulated within territorial states. The state survives and prospers only as long as it can hold together the territorial coalition of places that gives it geographical form. This involves pursuing policies and distributing resources in a deliberate way to maintain legitimacy. Under conditions of economic and political inequality between places, this will involve considerable geographical redistribution, especially when different political parties and powerful interest groups have different geographical bases (e.g., Blok 1974; Dulong 1978; MacLaughlin and Agnew 1986).

Fourth and finally, class, ethnic, and gender divisions have "national" histories as promulgated by political movements of one kind or another. Such divisions are often reified in political discourse. The extent to which they are accepted or viewed in terms of particular national (or in some cases international) discourses varies from place to place in terms of, for example, local shifts in work authority, economic dependence, local cultural forms, and the history of local experience with respect to particular social divisions (e.g., Joyce 1980; Jones 1983; J. Smith 1984; Tilly 1986; Andreucci and Pescarolo 1989).

Areal variation in political activity at any point in time, therefore, is not a product of *either* local attributes *or* general causes. Rather, there are contingent relationships between the activation of the causal powers of people, firms, and governments which "span" many places and the activation of those of people and organizations specific to them (Sernini 1988).

## Causation and Areal Variation

Some controversy and confusion has developed around realist views of areal variation such as this. In particular, it is alleged by some that, for such views, the local is a realm of contingency whereas the general (or supra-local) is the realm of necessity (e.g., N. Smith 1987). As contingency cannot be theorized, any focus on the local except as a direct function of the general (or macro) is a "return" to atheoretical empiricism. In fact, as Sayer (1988, 9) points out, this "involves a simple misunderstanding of necessity and contingency: just because the coexistence of A and B in a particular form is contingent, it does not follow that there is nothing necessary about how, once configured in this way, they interact." Making a distinction of the form local = contingent and general = necessity separates out locale and location from an integrated concept of place and privileges the latter (as necessity) at the expense of the former.

A related confusion involves fusing the abstract with the supra-local (or general) and the concrete with the local. Neither pairing is appropriate. Supra-local entities, for example, global corporations or political parties, are just as concrete as local firms or governments (Sayer 1988, 9). Conversely, the local and locality are abstract terms with complex meanings (Urry 1987, 441). More generally, realism offers no justification for the view that general causes are underway "first" and then "collide" with local contingencies to create the "concrete" or empirical.

To some other writers, a different perspective prevails: the local is seen as a separate and largely residual "effect." It is only present when there are "local cultures" or "neighborhood effects" that interfere with dominant "national" effects (e.g., Savage et al. 1987). "Locality" and national-level social categories form unintegrated poles of analysis. This viewpoint thus ignores location as defined previously and sees place as a product of local attributes alone (see Agnew 1987, ch. 6).

Finally, a focus on areal variation does not obviate the need to *theorize* about causal mechanisms and to attempt to identify them in empirical work by posing questions about how they relate to actual outcomes. The "danger of the empirical turn," as N. Smith (1987) identifies the problem, arises for him because of the lack of reliance on a single determination a priori. But it *could* also arise because in an openness to multiple, rather than single, determinations, explicit theorizing could decay and there would be a return, full circle, to Hartshorne's dilemma: how are the *significant* phenomena as expressed in regions or places to be explained? This is why attention to causes must parallel attention to outcomes.

This discussion of causes and outcomes in areal variation may seem well removed from the language and conventional interpretations of *The Nature*. In terms of its emphasis on dynamic or historical process, it is indeed, but there are a number of important continuities. First, this perspective shares with Hartshorne an emphasis on "phenomena" and how they come together as "element-complexes," rather than space as such. Second, there is a concomitant rejection of the privileging of particular spatial units as things-in-themselves. Third, as in *The Nature*, the distinctiveness of particular areas is not exhausted by "local" characteristics but includes links to "larger regions." Indeed, distinctiveness is a product of interaction between what is locally reproduced and wider worlds. Fourth, and perhaps most importantly from a philosophical point of view, the causal connections of interest to the geographer are, as Hartshorne first noted in *The Nature* (1939, 240) but restated more cogently later (1959, 18–19): "The mutual relationships among different phenomena at one place, and relationships or connections between phenomena at different places." This is plausibly similar to the view of causation characteristic of modern scientific realism and totally at variance with the Humean view of causality as a relation between *classes* of phenomena characteristic of logical empiricist/positivist philosophy of science.

## The Example of Italian Electoral Politics, 1947–87

The argument for areal variation can be illustrated using the example of the geography of Italian electoral politics since World War II. The purpose of this example is to emphasize both the major discontinuity with *The Nature*, the importance of the "phenomena" themselves and how their relationships change *geographically* over time, and the several continuities noted previously: a focus on "element-complexes," the absence of fetishizing or privileging particular spatial units, locality-"larger region" linkages, and "realist" causality. The geography of Italian electoral politics since 1947 can be characterized in terms of three distinctive political-geographical "regimes" that have dominated in different periods. This limited discussion draws on a number of more detailed sources including Agnew 1989; Brusa 1984, 1988; Corbetta and Parisi 1985; Corbetta et al. 1988; Mannheimer and Sani 1987; Rizzi 1986.

The first regime, dominant from 1947–63, is a clear regional (meso-scale) pattern of support for the major political parties based upon social-economic relationships that clustered regionally. The second, in effect from 1963–76, witnessed the expansion of the Communist Party (PCI) out of its regional stronghold into a competitive position with the Christian Democratic Party (DC). This expansion had different sets of causes in different places, but the net effect was to suggest a nationalization of the two major parties. The third, characteristic of the period since 1976, has seen increased support for minor parties, the geographical "retreat" of the PCI, and a more localized pattern of political expression in general, reflecting the increased "patchiness" of Italian economic growth and social change.

Rather than a fixed areal differentiation at a regional scale, the geography of

Italian electoral politics represents areal variation at a variety of scales. Different scales have dominated expression as place-based causal processes have brought about a changing balance of geographical sameness and difference.

## The Regionalizing Regime

The period 1947–63 is that of the "classical" electoral geography of Italy established most definitively by Galli and his colleagues (Galli and Prandi 1970). They divided the country into six zones on the basis of levels of support for the three major parties, the PCI, DC and the Socialists (PSI), and the strength of the major political sub-cultures, the socialist and the Catholic.

Zone I, the "Industrial Triangle," covers northwest Italy and includes Piemonte, Liguria and Lombardia. This is the region in which industrial production was concentrated before World War II and in which most new industrial investment was concentrated in the 1950s. Socialists, Christian Democrats, and Communists were all competitive in this region.

Zone II, "La zona bianca," covers northeast Italy and includes the provinces of Bergamo and Brescia in Lombardia, the province of Trento, the province of Udine, and all of Veneto except the province of Rovigo. The Christian Democrats were most strongly entrenched in this region and opposition was divided among a number of parties.

Zone III, "La zona rossa," covers central Italy and includes the provinces of Mantova, Rovigo and Viterbo; the whole of Emilia-Romagna except for the province of Piacenza; Toscana except for Lucca; Umbria; and the Marche, except for the province of Ascoli Piceno. In this region the PCI was most strongly established both in the countryside and the cities.

Zone IV, the South, includes the province of Ascoli Piceno, Lazio (except Viterbo), Campania, Abruzzo e Molise, Puglia, Basilicata, and Calabria. This zone is historically the poorest and most marked by clientelistic politics. In the 1950s, the Christian Democrats and the right-wing parties dominated the zone but were faced with increasingly strong challenges from the PCI and PSI.

The final two zones, V, Sicily, and VI, Sardinia, had more complex political alignments than the peninsula South. For example, the PCI was well-established in the southern provinces of Sicily (especially the sulphur mining areas) while Sardinia had a strong regionalist party.

There was a strongly rooted "cultural hegemony" in only two of these zones: "la zona bianca" and "la zona rossa" (Stern 1975; Muscarà 1987). But in electoral terms, support for specific political parties was remarkably clustered regionally in 1953 (Rizzi 1986): the PCI in the Center, the PNM (monarchists) and MSI (neo-Fascists) in the South and Sicily, DC in the Northeast and the South. In the 1950s, Italian politics followed a regional "regime" reflecting a similarity at the regional scale of place-based social, economic and political relationships.

## The Nationalizing Regime

The second period, 1963–76, marks a break with the regional pattern characteristic of the 1950s. Two electoral shifts were especially clear: the expansion of support for the PCI outside "la zona rossa" (along with its consolidation inside), particularly in the industrial Northwest and parts of the South, and the breakdown of "la zona bianca" as a number of small parties made inroads into the previously hegemonic support for DC in parts of the Northeast. The net effect of these changes was a *seeming* nationalization of the major parties, even though they still maintained traditional areas of strength.

These political changes were the fruit of the major economic and social changes that Italy underwent in the late 1950s and early 1960s. A major expansion occurred in manufacturing and industrial employment as a phenomenal boom or "economic miracle" drew the Italian economy away from its predominantly agrarian base. Whereas, in 1951, more than 40 percent of the active labor force had worked in agriculture, in 1961 the figure had fallen to 25 percent (13 percent in 1981). By no means all of this labor released from agriculture went into factories. A significant portion moved into the private service and public employment sectors. But manufacturing's share of GNP rose enormously, and there were massive migrations from the rural South to the industrial Northwest. Between 1955 and 1961, 50,000–90,000 people were added to the population of Milan annually, while the population of Turin rose by more than 50 percent in the 1950s, mainly in the latter half of the decade (King 1985).

At the same time that the industrial centers of the Northwest were experiencing such dramatic economic and social change as a result of the economic boom and massive immigration, the rest of the country was experiencing shock-waves emanating from the Northwest. The extreme South (Puglia, Basilicata and Calabria) was a major zone of emigration to the Northwest and, with the exception of Taranto, without much industry. Lazio, the region of Rome, received many immigrants as part of the general uprooting of rural and small-town populations. Industrial development projects in the South, especially in Sicily, designed to counter-balance the general northward movement of population, generated pockets of new social and economic relationships in the midst of a rapidly depopulating rural society. In all these places and among immigrants in the North, the PCI expanded its support in the late 1960s and early 1970s.

In addition to the geographical expansion of the PCI and its overall increasing share of the national vote, the other major feature of the period 1963–76 was the so-called "breakdown" of the Catholic sub-culture or hegemonic position in "la zona bianca" or Northeast and the loss of voters to DC subsequent to this. The argument is that DC, being largely an electoral rather than a mass party with a large membership, had relied heavily on affiliated organizations, many of a religious nature, to mobilize its support. But in the 1960s, as a result of heavy outmigration from rural areas in the Veneto, Trento and Friuli, the constituent regions of "la zona bianca," and the growing industrialization of some areas, such as Venice, Treviso, Trento and Pordenone, the traditional social networks and communal institutions upon which DC hegemony was based began to collapse (Parisi 1971; Sani 1977; Caciagli 1985; Chubb 1986).

The nationalizing political-geographical regime peaked in 1976 when DC and PCI together accounted for 73 percent of the national vote. Although this trend had distinctive causes relating to the geographically differentiated social and economic impacts of the "economic miracle" and their interplay with political and organizational traditions, it was widely interpreted as a permanent "nationalization" of political life (Agnew 1988b). DC and the PCI were now *national* political parties. The election of 1976 seemed to seal it once and for all as the PCI expanded in constituencies where it had previously been weak or where its previous support had stagnated (but not much in the Northeast): +10.6 percent in Naples-Caserta; +10.9 percent in Cagliari-Nuoro (Sardinia); +9.6 percent in Rome-Viterbo-Frosinone; and +9.2 percent in Turin-Novara-Vercelli.

## The Localizing Regime

The 1979 election indicated a much more complex geography of political strength and variation than had been characteristic previously. Since then all parties have been less regionalized than in the past (Rizzi 1986). The 1983 and 1987 elections suggest a trend towards a localization or increased differentiation of political expression. In 1983, DC lost 5.4 percent nationally. But the PCI was not the beneficiary. Rather, it was smaller parties such as the PSI and the Republicans (PRI) in the North and the MSI in the South that gained most. In 1987, DC recovered somewhat from 1983 but without a major geographical expansion. The major loser this time was the PCI, which not only lost ground in the Northeast and the Northwest but also in some provinces of "la zona rossa" to the PSI and a variety of smaller parties including the Radicals (PR), the "Greens," and Democrazia Proletaria (DP) (Leonardi 1987).

How can this localization be explained? One element is the increasingly differentiated pattern of economic change after a previous era of concentration. While the economic boom of the early 1960s concentrated economic growth increasingly in the Northwest, by the late 1960s there was considerable decentralization of industrial activity out of the Northwest and into the Northeast and the Center. This new pattern of differentiated economic growth led some commentators to write of the "three Italies"—a Northwest with a concentration of older heavy industries and large factory-scale production facilities, a Northeast-Center of small, family-based, export-oriented and component-producing firms, and a still largely underdeveloped South, reliant on government employment but with some of the "small-scale" development (e.g., in Naples and Caserta) characteristic of the "Third Italy" (Northeast-Center) (Bagnasco 1977). But this terminology, though useful as a general characterization of a new economic geography, masks what is in fact a much more uneven and differentiated pattern at a local scale. High concentrations of employment in major growth industries have, in fact, been widely scattered (Cooke and Pires 1985). There is no clear regional pattern. What is apparent is that there is no longer such a heavy concentration of industrial activity in the Northwest, even though that region remains generally dominant in terms of large factories and total output.

Other causes have also contributed to the contemporary localizing political-geographical trend. One of these has been the failure of parties to successfully adapt to recent social and economic change. In Trento and Udine, for example, DC has had problems adapting to the new economy. In large parts of the South and the Northwest, the PCI has been unable to capitalize on earlier successes mainly because, in the South, it has neither had control over the state resources that lubricate the politics of many parts of that region, nor been able to build a cultural hegemony. In the Northwest, its major "vanguard" of unionized workers has been much reduced in economic importance at the same time other parties have become better organized and the particular problems of southern immigrants have largely receded from the political agenda (Sassoon 1981; Caciagli 1985; Pasquino 1985).

The emergence of effective regional-level governments since 1970 has also reinforced the localization of interests and "sense of place." Where parties have achieved some strength and legitimacy through control over regional governments, they have been able to build local coalitions for national politics based upon the pursuit of local interests. The PCI, for example, has benefitted from its control of or participation in the regional governments of Emilia-Romagna, Toscana, and Umbria. But it has suffered elsewhere, and other parties such as DC or the PSI have benefitted, because of lack of control over patronage jobs and inability to write regional political agendas (Putnam et al. 1985).

Finally, to the extent that the former successes of DC and the PCI in, respectively, "la zona bianca" and "la zona rossa" rested on the "total" social institutions with which they were affiliated (unions, cooperatives, clubs, etc.), as well as social isolation, the shifting orientations of these institutions and the rise of the consumer society have opened up possibilities for the smaller parties. There is some evidence that, since the late 1960s, the ties between DC and the PCI and their supportive organizations, especially the unions, have weakened (Weitz 1975; Hellman 1987; Mershon 1987). The parties themselves are responsible for some of this. In order to expand nationally, they have often had to abandon or at least limit the ideological appeal that served so well in areas of traditional strength. They have also had to respond in some areas to "new" movements (e.g., the "Greens") which has opened them up for both factionalism and essentially localized forms of organization and ideology (Amyot 1981; Caciagli 1985). More generally, parties do not always "travel well." Thus, in comparing northeast with central Italy, the question of compatibility between party "style" and local "style" arises. Stern (1975, 223) notes:

> the evolution of two very different forms of political hegemony, each with distinct characteristics that necessitate sharply contrasting forms of maintenance. The Christian Democratic variety that flourishes in northeastern Italy is fueled efficiently by a stable social organization that deemphasizes the place of politics in community life. In comparison the Communist variant thriving in central Italy . . . accents the urgent attention that political matters should command among the local citizenry and thereby constantly reaffirms the relatively recent sense of legitimacy that underlies PCI control.

Of course, these hegemonies have always had local roots and, in some localities, their power is still quite visible, as recent studies of Bologna and Vicenza suggest (Kertzer 1980; Allum and Andrighetto 1982). There is persistence in place as well as change. At present, not only is support for the parties more obviously localized, so are the parties themselves.

This narrative and largely impressionistic account of successive "regimes" of political-geographical expression illustrates the conception of areal variation outlined earlier. In an analytical vein, the most obviously important causes identified in the narrative are shifting spatial divisions of labor, changing central government-territorial "patronage" relationships (especially at the regional scale), and shifts in political party ideology and organization, especially in relation to other organizations. As these have changed, so has the geography of political expression. But the relationship between cause and outcome is not a simple linear one. It is how the causes "come together" in diverse ways in different places with distinctive prior histories, and hence distinctive locale, location, and sense of place characteristics, that is important. Out of this interaction, geographical sameness and difference are produced. Areal variation is therefore best thought of as a dynamic process reflecting the differential operation of multiple causal mechanisms over space rather than either a mapping of static patterns at a regional scale (as in "traditional" political geography), explanation by a single national or global cause, or the identification of regional or local *effects* that are somehow "independent" of the *effects* of other scales (Agnew 1987).

## Conclusion

Fifty years after its first publication, *The Nature of Geography* can still provide a point of departure for a coherent geographical perspective, but only if it is interpreted as more complicated and fertile than the symbol it became in the 1960s. Above all, *The Nature* can be seen as providing pieces of an argument for areal variation. Hartshorne's lack of interest in science and "the phenomena themselves" fatally undermined the possibility of his overcoming the sameness-difference polarity that he seemed to recognize was a barrier to extending the scope of geographical understanding. Hartshorne wrote in a particular histor-ical-geographical context—perhaps we could call it a "timeless America"—in which fixity and meso-scale regions rather than change and place as defined here were dominant concepts of geography. It seemed important at that time to separate "geography" from the problems and interests of other fields. Hart-shorne's work reproduced this sense of reality.

The central claim of this paper is that *The Nature* can be considered more satisfactorily from a "realist" perspective, particularly in relation to its concept of causation but also more generally, than from the neo-Kantian and positivist readings that have predominated in the past. A number of continuities between *The Nature* and a realist perspective can be identified: a focus on element-complexes, an absence of privileging or fetishizing particular spatial units, locality-larger region linkages, and a commitment to a realist conception of causality.

There is also a major discontinuity: Hartshorne's lack of interest in "the phenomena themselves, their origins and processes" (p. 425) prevented the possibility of his defining the "significant" phenomena or subject-matter of geographical study, at least within the pages of *The Nature*. Hartshorne's "totalistic" conception of geographical knowledge prevented a problem-specific approach that would have been more amenable to the theorization of significance.

A realist theory of place, as applied here to the specific problem of the geography of political activity, provides one way of compensating for the lacunae in Hartshorne's argument. From this point of view, areal variation (the creation of sameness and difference) is seen as a dynamic process involving the operation of causal mechanisms (divisions of labor, etc.) that can produce different historical-geographical outcomes because of the differential activation of the causal powers of people and institutions embedded in particular places.

The apparently shifting geographical pattern of Italian electoral politics since World War II is used to empirically illustrate the continuity and discontinuity between "Hartshorne's" areal variation and that of a more coherent realist perspective. That Hartshorne was unable to provide a complete account of this latter perspective, to which his pieces of an argument for areal variation pointed, should not detract from the real intellectual contribution that can be found in *The Nature*, one obscured for so long by Hartshorne's own oracular style, easily subject to competing interpretations, and by the limited view of geography as areal differentiation that it has been widely seen as representing.

# References

**Agnew, J. A.** 1987. *Place and politics: The geographical mediation of state and society.* London: Allen and Unwin.

———. 1988a. Beyond core and periphery: The myth of regional political-economic restructuring and a new sectionalism in contemporary American politics. *Political Geography Quarterly* 7:127–39.

———. 1988b. "Better thieves than reds?" The nationalization thesis and the possibility of a geography of Italian politics. *Political Geography Quarterly* 7:307–21.

———. 1989. The geographical dynamics of Italian politics, 1947–1987. In *The geography of social change*, ed. L. Hochberg and C. Earle. Stanford, CA: Stanford University Press, forthcoming.

**Allum, P. A., and Andrighetto, T.** 1982. Elezioni e elettorato a Vicenza nel dopoguerra. *Quaderni di Sociologia* 30:355–97.

**Amyot, G.** 1981. *The Italian Communist Party: The crisis of the popular front strategy.* New York: St. Martin's Press.

**Andreucci, F., and Pescarolo, A., eds.** 1989. *Gli spazi del potere: Aree, regioni, stati: Le coordinate territoriali della storia contemporanea.* Florence: Usher.

**Bagnasco, A.** 1977. *Tre italie: La problematica territoriale dello sviluppo italiano.* Bologna: Il Mulino.

**Blok, A.** 1974. *The Mafia of a Sicilian village, 1860–1960.* New York: Harper and Row.

**Brusa, C.** 1984. *Geografia elettorale nell Italia del dopoguerra: Edizione aggiornata ai risultati delle elezioni politiche 1983.* Milan: Unicopli.

———. 1988. *Cambiamenti: Nella geografia elettorale italiana dopo le consultazioni politiche del 1987.* Milan: Unicopli.

**Caciagli, M. M.** 1985. Il resistibile declino della Democrazia Cristiana. In *Il sistema politico italiano,* ed. G. Pasquino, pp. 101–27. Bari: Laterza.

**Chubb, J.** 1986. The Christian Democratic party: Reviving or surviving? In *Italian politics: A review,* Vol. 1, ed. R. Leonardi and R. Y. Nanetti, pp. 69–86. London: Frances Pinter.

**Cooke, P., and Pires, A. da Rosa.** 1985. Productive decentralization in three European regions. *Environment and Planning* A 17:527–54.

**Corbetta, P. G., and Parisi, A.** 1985. Struttura e tipologia delle elezioni in Italia: 1946–1983. In *Il sistema politico italiano,* ed. G. Pasquino, pp. 33–73. Bari: Laterza.

———; ———; **and Schadee, H. M. A.** 1988. *Elezioni in Italia: Struttura e tipologia delle consultazioni politiche.* Bologna: Il Mulino.

**Cox, K. R., and Mair, A.** 1988. Locality and community in the politics of local economic development. *Annals of the Association of American Geographers* 78:307–25.

**Dulong, R.** 1978. *Les régions, l'état et la société locale.* Paris: Presses Universitaires de France.

**Entrikin, J. N.** 1981. Philosophical issues in the scientific study of regions. In *Geography and the urban environment,* Vol. 4, ed. D. T. Herbert and R. J. Johnston, pp. 1–27. Chichester, England: John Wiley.

———. 1985. Humanism, naturalism, and geographical thought. *Geographical Analysis* 17:243–47.

**Galli, G., and Prandi, A.** 1970. *Patterns of political participation in Italy.* New Haven, CT: Yale University Press.

**Gambi, L.** 1973. *Una geografia per la storia.* Turin: Einaudi.

**Gertler, M. A.** 1986. Discontinuities in regional development. *Society and Space* 4:71–84.

**Giddens, A.** 1979. *Central problems in social theory: Action, structure, and contradiction in social analysis.* London: Macmillan.

**Gould, P. R.** 1979. Geography 1957–1977: The Augean period. *Annals of the Association of American Geographers* 69:139–51.

**Gregory, D.** 1978. *Ideology, science and human geography.* London: Hutchinson.

**Guelke, L.** 1978. Geography and logical positivism. In *Geography and the Urban Environment,* Vol. 1, ed. D. T. Herbert and R. J. Johnston, pp. 35–61. Chichester, England: John Wiley.

**Harre, R., and Madden, E. G.** 1975. *Causal powers.* Oxford: Basil Blackwell.

**Hartshorne, R.** 1939. *The nature of geography: A critical survey of current thought in light of the past.* Lancaster, PA: Association of American Geographers [Reprinted with corrections 1961].

———. 1959. *Perspective on the nature of geography.* Chicago: Rand McNally.

**Harvey, D.** 1969. *Explanation in geography.* London: Edward Arnold.

**Hellman, J. A.** 1987. *Journeys among women: Feminism in five Italian cities.* New York: Oxford University Press.

**Johnston, R. J.** 1983. *Geography and geographers: Anglo-American human geography since 1945.* London: Edward Arnold.

**Jonas, A.** 1988. A new regional geography of localities? *Area* 20:101–10.

**Jones, G. S.** 1983. *Languages of class: Studies in English working-class history, 1832–1982.* Cambridge: Cambridge University Press.

**Joyce, P.** 1980. *Work, society and politics: The culture of the factory in late Victorian England.* Hassocks, England: Harvester Press.

**Kertzer, D. I.** 1980. *Comrades and Christians: Religion and political struggle in Communist Italy.* Cambridge: Cambridge University Press.

King, R. 1985. *The industrial geography of Italy.* New York: St. Martin's Press.

Leonardi, R. 1987. The changing balance: The rise of small parties in the 1983 elections. In *Italy at the polls, 1983: A study of the national elections,* ed. H. R. Penniman, pp. 100–19. Durham, NC: Duke University Press.

MacLaughlin, J. G., and Agnew, J. A. 1986. Hegemony and the regional question: The political geography of regional industrial policy in Northern Ireland, 1945–1972. *Annals of the Association of American Geographers* 76:247–61.

Mannheimer, R., and Sani, G. 1987. *Il mercato elettorale: Identikit dell'elettore italiano.* Bologna: Il Mulino.

Massey, D. 1984. *Spatial divisions of labour: Social structures and the geography of production.* London: Macmillan.

Meinig, D. W. 1988. Personal communication.

Mershon, C. A. 1987. Unions and politics in Italy. In *Italy at the polls, 1983: A study of the national elections,* ed. H. R. Penniman, pp. 120–45. Durham, NC: Duke University Press.

Muscarà, C. 1987. Dalla geografia elettorale alla geografia politica: Il caso italiano delle aree bianca e rossa. *Bollettino della Società Geografica Italiana (Roma),* Series 11, 4:269–302.

Outhwaite, W. 1987. *New philosophies of social science: Realism, hermeneutics and critical theory.* London: Macmillan.

Parisi, A. 1971. La matrice socio-religiosa del dissenso cattolico in Italia. *Il Mulino* 21:637–57.

Pasquino, G. 1985. Il partito comunista nel sistema politico italiano. In *Il sistema politico italiano,* ed. G. Pasquino, pp. 128–68. Bari: Laterza.

Pred, A. 1984. Place as historically contingent process: Structuration and the time-geography of becoming places. *Annals of the Association of American Geographers* 74:279–97.

Putnam, R., et al. 1985. Il rendimento dei governi regionali. In *Il sistema politico italiano,* ed. G. Pasquino, pp. 345–83. Bari: Laterza.

Rizzi, E. 1986. *Atlante geo-storico, 1946–1983: Le elezioni politiche e il parlamento nell'Italia repubblicana.* Milan: GSI.

Sack, R. D. 1974. Chorology and spatial analysis. *Annals of the Association of American Geographers* 64:439–52.

Sani, G. 1977. Le elezioni degli anni settanta: Terremoto o evoluzione? In *Continuità e mutamento elettorale in Italia,* ed. A. Parisi and G. Pasquino, pp. 67–102. Bologna: Il Mulino.

Sassoon, D. 1981. *The strategy of the Italian Communist Party.* New York: St. Martin's Press.

Savage, M.; Barlow, J.; Duncan, S.; and Saunders, P. 1987. Locality research: The Sussex programme on economic restructuring and change. *Quarterly Journal of Social Affairs* 3:27–51.

Sayer, R. A. 1984. *Method in social science: A realist approach.* London: Hutchinson.

———. 1988. "Meanwhile, back at the ranch . . . ." Problems of narrative and the new regional geography. Discussion Paper 100, Départment of Geography, Syracuse University, Syracuse, NY.

Schaefer, F. K. 1953. Exceptionalism in geography: A methodological examination. *Annals of the Association of American Geographers* 43:226–49.

Sernini, M. 1988. Aporie del localismo assoluto. *Archivio di Studi Urbani e Regionali* 31:113–65.

Smith, J. 1984. Labour tradition in Glasgow and Liverpool. *History Workshop* 17:32–56.

Smith, N. 1987. Dangers of the empirical turn: Some comments on the CURS initiative. *Antipode* 19:59–68.

Stern, A. 1975. Political legitimacy in local politics: The Communist Party in northeastern Italy. In *Communism in Italy and France*, ed. D. L. M. Blackmer and S. Tarrow, pp. 221–58. Princeton, NJ: Princeton University Press.

**Taylor, P. J.** 1988. World system analysis and regional geography. *The Professional Geographer* 40:259–65.

**Tilly, C.** 1986. *The contentious French.* Cambridge, MA: Harvard University Press.

**Urry, J.** 1987. Society, space, and locality. *Society and Space* 5:435–44.

**Warde, A.** 1988. Industrial restructuring, local politics, and the reproduction of labour power: Some theoretical considerations. *Society and Space* 6:75–95.

**Warf, B.** 1988. Regional transformation, everyday life, and Pacific Northwest lumber production. *Annals of the Association of American Geographers* 78:326–46.

**Weitz, P.** 1975. The CGIL and the PCI: From subordination to independent political force. In *Communism in Italy and France*, ed., D. L. M. Blackmer and S. Tarrow, pp. 541–71. Princeton, NJ: Princeton University Press.

# *The Nature*, in Light of the Present

ROBERT D. SACK

Department of Geography, University of Wisconsin–Madison,
Madison, WI 53706

T*he Nature of Geography* can be seen as a comprehensive, densely argued, and influential conception of the field, adding energy and ideas to a continuous process of geographical self-discovery. Viewing *The Nature* as part of a process rather than as its culmination is closer to the original intent of the book suggested by its full title, *The Nature of Geography, A Critical Survey of Current Thought in the Light of the Past.*

Light from the past illuminates Hartshorne's work, but the rays travel in the other direction too—the past is illuminated by the present. Both directions will be employed here, but I will concentrate mostly on the second. I will discuss how present geographic debates are themselves understood best as attempts to weave together elements from an extremely broad and diverse "intellectual terrain." *The Nature* also attempts to negotiate this terrain but draws upon relatively few threads from each part, when compared to the numbers involved in current geographic thought. The selection of a few threads to interweave allows Hartshorne to achieve a tight set of logical interconnections. This tightness of logic accounts for the often-noted inward-looking quality of *The Nature*, even more than does its historicist focus mostly on the works of German and Anglo-Saxon geographers.

*The Nature* will be discussed in light of the current intellectual terrain with which geography and other disciplines are attempting to grapple. This terrain has two levels. One involves intellectual methods or "perspectives" and has an epistemological quality to it. Academic disciplines can view the world from the perspectives of science, the natural and social sciences, the arts, and so on. *The Nature* stakes out a particular place on this level by employing a scientific perspective, though not quite like the nomothetic views outlined in conventional philosophies of science. The exact qualities of this perspective have been the cause of considerable puzzlement for the profession.

Perspectives provide a means of viewing things, and among the most basic things to view are the fundamental forces that we think exist in the world, and which most affect us. These form the second level of our terrain—an ontological level of "forces." The two levels are interrelated: perspectives are selected in order to bring certain things into focus, and the things viewed bring content to the perspectives. In philosophical terms, every epistemology presupposes an ontology, and every ontology presupposes an epistemology. Thus, the levels are basic components in serious intellectual discussions, as can be seen in *The Nature*, and will form the basic context for examining Hartshorne's book.

This paper is divided into three sections. The first introduces the intellectual terrain of perspectives and forces. The discussion of perspectives will consider the range from "somewhere" to "nowhere" and the location of chorology within this range. In the discussion of forces, I will argue that modern intellectual discourse recognizes four major realms: nature, meaning, social relations, and human agency. Much of current geographic debate concerns their character and their connections (Sack 1988).

The second section will discuss how *The Nature* can be seen to weave elements from this intellectual terrain. *The Nature* of course, does not mention these particular terms, or levels, nor do these issues form part of the basic organization of the book. Hartshorne did not write with these realms and levels specifically in mind; nevertheless we can interpret his work as addressing them in very illuminating ways. I will argue that, in Hartshorne's work, the concepts of place, including the generic and specific place, stand as the geographical interconnection of the four realms: they incorporate elements from meaning, nature, social relations, and human agency. Furthermore, I will argue that for Hartshorne, the specific place is central to the field of geography and comes to the fore when viewed from a particularly important geographical perspective that he calls chorology. Chorology then is the geographic perspective; specific place is the geographical object that it draws into focus. One of the puzzlements that Hartshorne has left us with is deciding if the specific place can be identified from any other perspective—if it is really out there.

In the third section, I will consider the reactions that *The Nature* helped set in motion. I will argue that current geographic debate draws on a far more complex set of elements from this intellectual terrain than does *The Nature*. In this respect, geography has become a microcosm of the intellectual world, and virtually nothing on this terrain is allowed to remain constant or fixed. In this kaleidoscope of issues, *The Nature* seems somewhat constraining, and geographers search for a more comprehensive synthesis.

# Intellectual Terrain

## Perspectives

A basic property of human consciousness is our capacity to see the world from a particular point of view and also to see the world from any number of more distant perspectives that can contain this and the other particular views as special cases. At this very moment, I can describe my world as the particular setting or context in which I am located and through which I am acting. It contains my office, my books, my department, and so on ... and out of my window I can see others who are near my office. My view also contains my opinions and judgments about what I see in my world and what it means. A view from somewhere else would involve the ability on my part to see myself in my own office looking out of my window and holding the opinions I hold. Others also seek to see themselves in their world from outside themselves. Both commonsense and the scholarly literature on child development claim that gaining this intellectual detachment and perspective on oneself is a mark of maturity. It is also a move from subjective to increasingly objective positions, when these perspectives can be shared by others. In this regard, objectivity is based on intersubjectivity. But it is more than that if we, as realists, believe there really is something out there to see, independent of our view. In the words of Thomas Nagel:

> to acquire a more objective understanding of some aspect of life or the world, we step back from our initial view of it and form a new conception which has that view and its relation to the world as its object. In other words, we place ourselves in the world that is to be understood. The old view then comes to be regarded as an appearance, more subjective than the new view, and correctable or confirmable by reference to it. The process can be repeated, yielding a still more objective conception (Nagel 1986, 4).

Nagel describes seeing the world from a particular perspective as a view from somewhere, and viewing the world from a more distant perspective as tending toward a view from anywhere or nowhere. There are many paths to a view from nowhere. Art is one of these, but the most prized and general in our culture is science. Nagel makes the extremely important point that while we can shift from one view to another and develop ever more distant and objective perspectives, we cannot develop a single perspective, even through science, that itself contains both end points. It is impossible, in Nagel's words, "to combine the perspective of a particular person inside the world with an objective view of that same world, the person and his viewpoint included" (Nagel 1986, 3). Another way of stating this is that, even when we mentally travel outside ourselves and see us and the world from distant objective perspectives, we still have one foot left, so to speak, in our own particular place or context. This is an inescapable part of being human.

The cleavage between seeing the world from somewhere and nowhere does not simply happen by force of mental powers alone. (If it did, we would be explaining a view, or an epistemological position, by a single realm of force or ontological structure.) As is well known, both our own personal actions and our social structures are important factors in the development of perspectives. Child development research is dedicated to describing how our very manipulation of the world gives rise to an awareness of ourselves as agents in the world (Piaget and Inhelder 1956). In a vast, highly specialized, interconnected, yet constantly changing world, we must build on those experiences which we do share, perhaps by virtue of mass culture, and examine them in as public and objective a light as possible.

In our society, the most direct and unambiguous language for public discourse is that of science. Science is really a range of approaches that employ discursive languages. These languages provide some common ground from which to view at least the non-human world of nature. Science is also one of the preferred perspectives for public discourse about people, their minds and their behavior. Whether it can convey the facts and meaning of human existence is now hotly debated, but it is not an issue we will discuss here, except to say that if the general perspectives of science cannot illuminate the truth about human nature (and if our insistence on using it results only in scientism), there is as yet no other perspective our society would be willing to accept in its place.

Thus far we have said that perspectives can range along several paths from somewhere to virtually nowhere. But what are examples of these perspectives and what are their particular qualities? What kinds of things do they help us see? The potential range of views that can be of use for answering geographic questions have been examined elsewhere (Sack 1980), but the actual professional field of geography works within a subset of this range. Professional geographers do not evoke place through poetry or painting or other views along the aesthetic path, although as Tuan (1989) shows, aesthetics are a basic facet of all of our perceptions and actions. Rather geographers use discursive forms of representation which can be thought of as broadly scientific. At the great risk of over-simplification, it may be useful to array the geographical perspectives along a continuum. At one end are the various "humanistic" perspectives on place. These tend not to develop a very technical language, but rather to use more ordinary, everyday expressions so that the richness of the experience of being in place can be captured. These perspectives are closer to a view from some-where, though they actually hover above the somewhere in search for common themes that appear in all "somewheres"—in the shared human experiences of being in place. From these perspectives, place is seen as a complex amalgamation of features—a field of care—and space becomes a thinned-out series of places (Tuan 1977). At the other, more distant, end are the spatial analytic perspectives of positivism which see the geographical world as a space and the particular places or "somewheres" simply as the locations of attributes in that space. These perspectives tend to use technical terms and quantification in order to find generalizations and to simplify human experiences so that they can be gener-alized and communicated as quickly as possible. The perspective taken in *The Nature* is complex and has given geographers considerable difficulty, but it can be said to be closer to the spatial-analytic than to any other. Hartshorne asserts that geography is a science and that it should view its phenomena objectively. Yet it is not exactly a generalizing or nomothetic science because its principle method is chorology. The difficulty arises in the meaning of the chorological perspective and its relationship to nomothetic science. These issues will form the topic of the second section.

## Forces: Nature, Meaning, Social Relations and Agency

Even though the preceding subsection was entitled perspectives, we could not avoid mentioning forces that might contribute to the development of particular perspectives. This is because perspectives and forces are inextricably interconnected. Here, though, we will concentrate on the fundamental forces that the natural and social sciences, through their theories, point to as existing out there and influencing human behavior. What then are the major forces affecting our lives? Contemporary answers seem to draw on the four loci or realms of power mentioned earlier: nature, meaning, social relations, and agency. These realms are interdependent and help constitute each other. All of the great social theories recognize this, but none draws equally from the four. Each theory emphasizes the processes of a single realm. Our review is not intended to evaluate the claims of these theories, but rather to describe the kinds of forces they claim exist.

*Nature.* Few terms have as many different meanings as does nature. Geographers have used it in a number of ways, ranging from scenery to something untouched by humans. Here the realm of nature refers to the basic forces, including elements and relations that are identified by the natural sciences. These involve mass, heat, light, gravity, magneticism, electricity, chemicals and chemical reactions, cells, genes, organisms, and biological growth and decay. People are also part of the natural world in that they too are composed of these properties and affected by these forces. In a sense, people form a subsystem within the larger system of nature.

Several important claims have been made to the effect that humans are not only composed of natural elements, but that their behavior can be understood in terms of natural forces. This is what we are told by the theories of sociobiology, of neurophysiology, and the various geographic theories of environmental determinism or environmentalism, and by facets of human ecology. These theories see natural forces, in the form of energy, genetics, climate, etc., as molding our social organizations, and even our values and ideas. These theories contain room for humans to alter parts of nature, but only as a subsystem would to a general system. We do not change nature's laws nor redirect its major forces. On the contrary, nature molds our behavior. At the time of *The Nature,* environmental determinism and its weaker variants (see environmentalism in *The Nature,* 100, 122, and 124–25) were important views, and Hartshorne hoped to prevent them from overwhelming geography. In *The Nature,* Hartshorne argues against having the natural world and its forces become the "geographic factor[s]" (122), and he does not think it prudent, or historically accurate, to define the field as the study of the relationship between natural and human forces. The issue is discussed within the general context of geography as a science of relationships (1939, 120–27). In other words, he wants to make room for the existence of other kinds of forces and allow them some autonomy.

*Meaning.* Another set of answers draws on the realm of meaning and the mind. Here prominence is given to the power of ideas, concepts, and symbols. This perspective claims that our actions are driven by meaning. Just as the realm of nature provides theories that purport to explain virtually everything (totalizing theories, as some have put it), so too can theories about the mind. Freud (1952) argues that basic psychological tensions create social organizations which also transform nature. Lévi-Strauss (1966) argues that basic oppositional characteristics of thought "structure" society and our conceptions of the natural order. Historians and philosophers of science remind us that what we know of nature comes from our natural science models and hypotheses. These are languages that talk about and represent nature. Indeed, we cannot know nature or anything else without representing it in thought. To this extent, the mind fashions the natural world. Some might even say that the natural world is a mental construct. This does not need to lead to the claim that there is no world out there independent of the mind. It could simply mean what it says: that we cannot escape knowing the world through our own sets of attitudes and beliefs. That is, we construe the world and cannot know it as it really is.

Believing that there is a world independent of the mind, though we may never know it as it really is, is one form of realism. Doubting its existence because we cannot know it as it really is, but only as we construe it, is a form of idealism. The spark set off from the irreconcilable tensions between the two is skepticism. Some might believe that recognizing the power of the mind, viz., our thoughts and symbols, in affecting our behavior, commits us to idealism, but this is not the case. Theories from this realm can be idealist or realist. In fact, all of the theories I have mentioned in the realm of meaning are realist theories. Freud, Lévi-Strauss and many others argue not only that there is a world independent of the mind, but that the power of the mind is real—to the point that the structures may actually be wired into the brain, so to speak. It is certainly not an epiphenomenon of something else, like nature, nor is it a "superstructure" of some base, like capitalist social relations.

Meaning as a force can easily be confused with meaning as a perspective. Both involve the mind. A perspective like science is a way of seeing things. It can, however, be thought of as a force, as when we say that "science helped create technology," or that "the scientific perspective supports economic growth." Often the literature is unclear about whether it is considering a mental property to be a force or a perspective. This is particularly true in the use of the term culture. The meaning of culture is vague, but the most important of its connotations refers to a people's ideas, values, and beliefs that can be manifested in various forms, including material artifacts: in short, to a people's symbols (Thrift 1989, 14). Culture, in this general sense, has often been thought of as a prism or lens—a perspective—through which people view the world; a people cannot escape their own cultural biases. But culture has also been used to mean a force, as when we discuss the cultural transformations of nature, or when cultural concepts are incorporated as variables in cultural ecology studies. As we shall see, Hartshorne considers certain mental properties as both forces and perspectives. On the one hand, he sees the chorological method as a perspective to investigate what is out there, and yet, what it sees is the specific place, which might not be able to exist except as a mental construct.

*Social Relations.* A third set of answers considers the primary forces to emanate from social relations. These relations concern the properties of society, economy, and politics. Marxist theory makes the most comprehensive claims in this realm.

For many Marxists, the realm of meaning is "in the last instance" engendered by the forms of social relations. Social relations also transform the realm of nature, creating a "second nature" (Smith 1984). Cultural Marxists, though, have attempted to loosen the hold of social relations by introducing culture or the realm of meaning as a force that is virtually coequal to that of social relations (Cosgrove 1984; Cosgrove and Daniels 1988; Daniels forthcoming; Jackson 1988).

Other, less sweeping social theories also see the principal force molding meaning and nature to come from elements of social relations, but not necessarily from class structure. Some theories, for example, attempt to explain our ideas and values in terms of our particular incomes and statuses, and our images and uses of nature as replicas of our own social organizations. *The Nature* discusses social relations, but in terms of how they create areal differentiation.

Nature, meaning, and social relations, then, are loci of a contemporary intellectual terrain. The realms are vast and help constitute each other, and whereas the theories within each share a common perspective about the fundamental forces that shape human nature, these same theories often work these forces in different and mutually contradictory ways. Theories from each realm can both draw from those of other realms and yet claim their realm to be primary. For example, Freud (1952) views the mind as connected to natural or biological drives, but his theory itself concerns the characteristics of meaning and not of biology. Similar interpenetrations among the realms can be seen in the works of Lévi-Strauss (1966) and Marx (Smith 1984). For the latter, man is first a part of nature, but then finally dominates nature.

Each realm offers plausible cases for its primacy and for its power to influence the others. But it is important to point out that the three realms are at an intellectual stand-off. The deadlock becomes even more evident when we add to the competing forces the complexity of selecting perspectives.

Philosophical systems do not help to reduce the deadlock. We noted that, at first blush, the realm of meaning might appear to imply a form of idealism. But, to Freud, Lévi-Strauss, and others, mental properties are "out there" in the real world. Furthermore, although we may never be able to know the world except through the prisms of our minds, this does not have to mean that there is no reality independent of the mind. Similarly, it might appear that physical science theories must be realist, but they can equally well imply a form of idealism, as when a physical scientist may believe that nature is only what her theories tell her.

*Agency.* The three loci differ in their view of forces that control or shape human behavior, but they share the perspective that such forces do exert control. This is, of course, the premise upon which any natural or social science rests. Our fourth area arises when we remember first that individuals are agents who create and sustain meaning and social relations (that is, forces cannot operate except through us); and second, that we might, in some sense, be free to create, or to make some of our decisions on the basis of our own free wills. This does not mean the absence of all constraints on our behavior, but it does mean that (some of) our behavior is free. This realm of freedom is extremely complex and ambiguous. Some may contend that parts of it have recently been discussed under the concept of agency (Giddens 1984). Agency, though, is an unfortunate term to use because its primary meaning is exactly the opposite of being free; an agent normally means an instrument or vehicle conveying some force. Giddens's use of the term emphasizes the important point that agents not only convey forces but that, in so doing, they actually sustain and modify these very forces. Yet what freedom might exist in Giddens's agents is illusory and stems from the fact that some forces affecting the agent may be unspecified and at a scale that is different from the one being analyzed by a particular set of theories. The crucial point is that a human agent becomes a true force only when she or he is free, otherwise one is an instrument, albeit a necessary one.

Most, if not all, of us believe that we do have free will; not everything we do is determined for us by outside forces, but we encounter problems if we further articulate this belief. It seems as though the existence of free will is predicated on the perspective we take. Once we see our actions from outside of ourselves and attempt to explain them, freedom seems to slip away. Suppose that I could choose between two desserts: chocolate ice cream or an apple. I think it over, tempted by the sugar and taste of chocolate, and by the nutritional benefits of the apple, and finally I choose the ice cream. I had a choice, and it was I who selected one over the other. But why did I choose ice cream (Nagel 1986, 110–18; 1987, 47–58)? If I say that my selection was based on an overpowering urge for sweets or self-destruction, I am saying that I really did not have freedom because my choice was in fact determined by these urges. The only way I can retain my sense of freedom is if I claim that the action was caused by me alone—by my will, and nothing else. But in that case, what exactly am I? A will? And an uncaused one at that? Is saying that I caused something to happen equivalent to saying that the happening was uncaused? The problem is that any explanation of my choice, and any attempt to objectify it and see it from somewhere else, will push free will away in favor of some cause or force that determined my action, and make the belief in free will an epiphenomenon.

How the four realms of forces are connected is at the core of modern social theory. Yet few of the social sciences concern themselves with the entire sweep of issues because they do not normally examine the relationship between people and nature. This is where geography's focus makes it central to the debate. Geographers are interested in all four realms and a powerful locus of the four is geographical place. Place, as we well know, has many meanings. Even if we do not settle on a single comprehensive definition, we still can see how many of the ordinary activities that we associate with being in, or making, places involve components of nature, meaning, social relations, and agency. Not all places, of course, need to involve the four equally. Just as there is no consensus within social theory about how the four are interrelated, there is no consensus among geographic theories about the mix of the four and the role of place in shaping their connections.

It is amidst this conceptual terrain of forces and perspectives then, that much geographic heat, if not light, has been shed. Let us shed some of this light on

*The Nature* and consider how it stands on these issues. Though *The Nature* was not written to address them specifically, in retrospect it does in fact encompass them in a tightly woven framework.

# *The Nature*

We will divide our discussion of *The Nature* into three parts. The first concerns areal differentiation and the specific place. (Hartshorne uses place, area, and region interchangeably; we will refer to them mostly by place.) The second is about how place draws together the four realms of forces. The third examines the chorological perspective that Hartshorne claims is necessary to analyze the specific place. Before we plunge into these three themes, we should be alerted to several interesting undertones. One is that while Hartshorne discusses how place comes about and embodies elements of the four realms, he does not raise place to a force in itself or examine how places (or space) affect processes, except insofar as they affect other places. Another is that he is in fact ambivalent about the existential character of place. Does it really exist or is it a mental construct? As we shall see, this ambivalence arises from a most complex and interesting tension in Hartshorne's synthesis between his use of the mental as a force or power in constituting specific place and his use of the mental as a perspective for viewing place.

## Element Complexes and the Specific Place

Hartshorne sees geography embedded in the fundamental interconnectedness of the world: the "unity of all nature" (Hartshorne 1939, 68, paraphrasing Humboldt and Ritter). This unity is expressed through place, area, or region. Places, areas, or regions reflect the fact that human and natural activities are distributed differently over the earth. This areal differentiation, which Hartshorne later changed to the clearer term variation (1939, 475, note 35), and which we will employ, is a basic characteristic of reality, and people have been curious about it from the very beginning (Hartshorne 1939, 130). Curiosity about places is, then, a particular form of curiosity about the world.

> Obviously there are many different ways of studying the world, but since men as individuals, and as individual groups, do not in any very full sense live in the entire world, but each in a relatively restricted area of the world, one of the most significant methods of studying the world is to study it by areas (Hartshorne 1939, 131).

The geographer's task is to describe and explain this variation and construct a picture of the world based on its specific places. "Geography, in brief, can demand serious attention if it strives to provide complete, accurate, and organized knowledge to satisfy man's curiosity about how things differ in the different parts of the world . . ." (Hartshorne 1939, 131). Geography is thus "concerned with studying the areal . . . [variation] of the world" (Hartshorne 1939, 133). It "analyzes how the most heterogeneous materials of areas are joined

together by causal relationships to form the character of the different areas of
the world, and of the world as a whole" (Hartshorne 1959, 30). How important
is this perspective? We can "assume that whatever has engaged intellectual
curiosity of peoples of all times—including many of the most eminent thinkers—is a subject worthy of advanced study" (Hartshorne 1939, 130).

Place (or area or region) is central to experience and its capacity to integrate
phenomena is unbounded. This is reflected in the claim by Hartshorne that
there really are no particularly geographical facts. Any element from any realm
can be a part of place and be geographical.

> Geography does not claim any particular phenomena as distinctly its own, but rather
> studies all phenomena that are significantly integrated in the areas which it studies,
> regardless of the fact that those phenomena may be of concern to other students
> from a different point of view (Hartshorne 1939, 372).

This interest in place is geography's distinct and enduring contribution. As
Sauer (quoted by Hartshorne 1939, 130) puts it:

> No other subject has preempted the study of area.... If one were to establish a
> different discipline under the name of geography, the interest in the study of areas
> would not be destroyed thereby. The subject existed long before the name was
> coined.... The universality and persistence of the chorologic interest and the prior-
> ity of claim which geography has to this field are the evidences on which the case
> for the popular definition may rest.

But what exactly are places? This is a complex problem, and from the very
beginning, Hartshorne attempts to balance the view that place is a naively given
and real element in nature with the view that place needs a particular kind of
perspective to sustain it. The tensions between the two positions emerge if we
move from the apparently concrete character of element complexes to the less
concrete bounding of these complexes to form places. Let us begin with the
element complexes.

Hartshorne's meaning of place rests on the existence of a public and objective
system of location that describes the uneven distribution of things and forces.
Unevenness, then, is a combination of facts and uneven causal interrelationships
among unevenly distributed facts. To use Hartshorne's own example (1939, 415),
rainfall and corn yield are not only uneven in their distribution, but are uneven
in their functional or causal interconnectedness over space. In one location,
corn can be extremely responsive to slight variation in precipitation, and in
another, it may not. Hartshorne uses the term "element complexes" (1939, 428–
31 esp.) to identify such uneven functional interdependencies, which are part
of the general interconnectedness of things in the world.

Place, area, or region, in this view is not much more than the circumscription
of these interdependences: place, area, or region is the areal extent of element
complexes. If the dependencies are general kinds and form the same associations,
then the places are generic. If the dependencies are uneven over space, then
the places are specific. "When we speak of the functions of areas, we are not to
forget that in reality the area is not a thing that functions, it is only certain

things within it that have functional relations to things in other areas" (1939, 445). And when we speak about interconnections among places we are really speaking about "some of the elements of one area and some in another" (Hartshorne 1939, 396). Thus place is really a type of whole or system.

Yet Hartshorne is well aware that even if places mean only drawing lines around things, we soon begin to think in terms of places, and these take on a reality of their own. Many of these have names, and many others are instituted by political or social authorities, the most familiar cases being political territories. We attribute powers to such places, as when we say "New York City attracts tourists," or when we say "you're not allowed to do this 'here' but you may do it 'there'." If we did not perceive the world in terms of places, then geography would be the naive science of element complexes rather than of places. For Hartshorne then, element complexes seem to have a somewhat firmer foundation in reality than do places, because places require some form of intellectual effort to construct. Once we shift our attention from element complex to place, its existential character becomes even shakier and inevitably invokes questions of scientific perspective.

Hartshorne claims that the facts of areal variation could be constructed by nature, free agents or social forces, but they are not products of the mind. Once they are constructed, they are simply there—naively given. Science, as the accurate and orderly description and analysis of the world, will provide a transparent lens through which to view these facts. Geography then investigates "the relationships grounded in nature itself" (Hartshorne 1939, 72); and "[s]o far as pure description is concerned, the method of the geographer is photographic in character, with the distinctive personal reactions of the observer reduced to a minimum" (1939, 133). But Hartshorne does not have the same confidence about the transparent quality of a scientific perspective when trained on the specific place.

Not all of the world that the geographer sees is really out there because places, whether specific or generic, do not exist by themselves but require mental support.

> Since nature (reality) has been so unkind as not to present us with obviously individual concrete objects . . . we must construct our own, by intellectual activity. . . . [I]t follows that any principles we attempt to develop can have no more validity than the 'objects' we have constructed as their foundation (Hartshorne 1939, 254; see also 395).

Places, then, do not really affect each other. Rather it is the phenomena in place that affect other phenomena. In other words:

> The area itself is not a phenomenon, any more than a period of history is a phenomenon; it is only an intellectual framework of phenomena, an abstract concept which does not exist in reality. . . . Indeed we cannot properly speak of relationships (other than purely geometric) between areas, but only between certain phenomena within areas. Likewise, the area, in itself, is related to the phenomena within it, only in that it contains them in such and such locations (1939, 395).

Place then, whether generic or specific, appears to be as much about points of view as about reflections of the "facts" (Hartshorne 1939, 380, 457). The curious point here is that Hartshorne is not concerned with the role of the mind in all perception; thus he does not call into question whether science is indeed ever photographic (1939, 133) and whether all facts, including those of element complexes, could be naively given. But perhaps this is not such a puzzle if we consider that, with few exceptions, the accepted philosophical wisdom of the 1930s was that science is a "transparent" lens through which to view nature as it really is. The surprise, then, is that Hartshorne doubts that this transparency applies to place. One reason for this doubt may lie in the general problem of conceiving of space and place as forces in themselves. Another reason is that Hartshorne wants to refute the ideas of natural regions and regions as organisms. What better way to accomplish this than to show that place is not all that natural? But doubts about the existential quality of place are driven most by a concern with the role of the mind in perceiving place. Hartshorne believes that perhaps places in general, but most certainly the specific place in particular, are visible only through a particular perspective, and this raises the possibility that the perspective itself helps constitute the object. The perspective then becomes a force. But before turning to the perspective, we must consider the question of how element complexes and places involve all four realms.

## Places and Forces

*Nature.* In one sense, nature sets the stage for the entire work and then takes a more passive role. Much of what the forces of nature produce is areally differentiated. Water and land, landforms, climate, soils, vegetation, and animal life are all distributed unevenly over the globe, and each type of distribution is related to the other. Geography examines the "well-known fact that things are different in different areas of the world and that these variations are somehow causally related to each other" (Hartshorne 1939, viii).

Whereas the uneven distribution of elements of the natural world have had an enormous impact on human actions, nature does not become the geographic motor so to speak; Hartshorne takes strong issue with environmental determinism and environmentalism. He rejects the concept of natural regions and argues that a discipline cannot be based on studying a specific set of causal relations such as the effects of nature on behavior or the "ecological" interconnections between humans and nature (1939, 120, 126). Nature is placed in the background for two reasons. One is that human forces are also important in understanding the areal differentiation of the world. (Hartshorne [1939, 43] cites Kant as including humankind as one of the five principal agents affecting changes on the earth.) The other is that areal differentiation and especially the specific place require a point of view to see and to be sustained.

*Social Relations.* Hartshorne emphasizes society's active role in transforming nature and affecting element complexes. Political territory is a prime example of places being created by social relations. "[T]he state . . . is not only a social organization, it is at the same time the organization by man of a particular section of the earth's surface . . ." (1939, 243). Economic forces are another element of social relations that play an important role in molding element complexes and landscapes. This is because

> for most of the people of the world, a major part of their activity is concerned with ways and means of keeping body and soul together, or more correctly, with making a living—i.e., economic activities. Likewise the greater part of the land area of the world used by man is used for economic activities. These economic activities, particularly those involving the use of the greatest amount of land, show marked areal differentiation in different parts of the world (1939, 334–35).

As for the relative importance of the particular forces of social relations, Hartshorne assigns the greatest weight to the economic: "in the complex of cultural phenomena associated with economic activities, we have the single group of phenomena of greatest importance to cultural geography . . ." (1939, 335). What then is the most important type of economic force? Here, again, the answer does not lie in some abstract theory, but rather in the extent to which particular forces mold the landscape. Measuring the importance of forces by their effects on landscape could result in a very different picture of relative weights of forces than would come from a social science that considers the significance of forces by their effects on factors such as social stability or poverty. Using landscape change as the criterion, Hartshorne argues that agriculture is the leading economic factor transforming place. Yet, in terms of economic theories, agriculture, as well as other sectors of the economy such as manufacturing, may not always be the basic units. Rather theories could instead point to raw material, labor, capital, and types of economic systems. For Hartshorne, though, (who discusses types of economy but does not dwell on them; 1939, esp. 400–02) the importance of a process is judged by its contribution to areal differentiation, and if this means that the process receives a different weight than it would from the other sciences, then so be it. (Note that physical geographers do not always have to plumb the depths of physical theories to explain landscape changes.)

We have considered how social forces create areal differentiation. But Hartshorne also alludes to the possibility of social forces leading in the opposite direction by creating geographical homogenization. "[I]f the area is under the control of man, its landscape cover is ordered and arranged in definite units each of which is strikingly homogeneous and sharply separated from the others" (1939, 172).

Many contemporary geographers have assumed that homogenization, with functionally useful local variation, is the primary effect of modern world culture and economy (Harvey 1982, 1985). But Hartshorne resists going this far, not because he doubts the power of social forces, but because he (mistakenly) believes that such homogenization would have to be consciously planned:

> Though man has thus produced landscapes which approach mosaics, and has created organic space units of farms and towns, and could, no doubt, produce larger units

in regions if he chose, this latter he has not in any real sense attempted to accomplish. Except for the limited degree represented by his planned and organized routes of transportation, man does not, either consciously or unconsciously, organize a region as a unit (1939, 280).

*Agency.* The realm that we have called human agency not only refers to the individual and his/her actions, but to his/her free will. Hartshorne makes only brief mention of the role of individual action in producing and maintaining places (see Hartshorne 1939, 279) and does not mention free will as a force (though Hartshorne 1959, 153–56, addresses it explicitly). Yet it is clear throughout his discussion of the limitations of the nomothetic method in geography that he believes our freedom to make choices will thwart the efforts of a nomothetic social science. This argument for the role of free agents appears, not directly in his discussion of landscape formation, but in his discussion of the geographic perspective or method of study.

*Meaning.* In Hartshorne's work, the realm of meaning is the point of contact between forces and perspectives. It is the use of chorology as both a force and a perspective that puts the existential character of place in doubt. But before we turn to the chorological viewpoint, we should note that Hartshorne employs meaning as a force in other ways as well. In particular, he argues that geography should not search for meaning in landscape only by examining the most visible manifestations of a people's attitudes and beliefs such as their church buildings and their capitals and courthouses, but rather consider the distribution of the less tangible meanings themselves (1939, 199–235). Indeed he is very much interested in the geography of the less visible powers of values, attitudes and beliefs, if they in turn could be shown to contribute to areal differentiation.

## The Chorological Perspective

The perspective Hartshorne advocates is chorology, and it is part of the general scientific view. By science Hartshorne means the objective, accurate and orderly description and analysis of the world and its parts. Hartshorne assumes there would be very little controversy over the descriptive part of science because facts, he believes, are out there and speak for themselves. The difficulty comes with analysis. Most of science seeks to find generalizations or laws that could explain and predict (the symmetry of course is questionable in open systems). It is at this point, according to Hartshorne, that regular science cannot capture the rich qualities of place. The key is that the specific place does not fall within the scope of generalizations. Geography needs something else. Let us pursue this point, but in terms of element complexes.

Element complexes of interest to geography can often be found again and again in the landscape. Examining these kinds of associations could lead to generalizations, and the types of places these associations produce would be called generic places. Thus we can expect, in general, to find certain kinds of vegetation associated with certain soil and climate types, and these associations lead to places of a general kind. But the most geographically interesting associations, according to Hartshorne, are those that differ from place to place. These still result from causal laws (as the corn yield example illustrates, 1939, 415), but they are exceptional associations—ones that generalizations could not encompass without extreme oversimplification. Geography, of course, studies both generic and specific places, but it is the latter that truly represents its mission. These exceptional associations give rise to specific or unique places and differ from the generic kinds of places that one would expect to occur more than once.

Hartshorne sees that geography would not be distinct from other sciences if it sought only to examine generic types of places. This is because all generalizations or laws (and these are what the other sciences attempt to discover) explain (potentially) repeatable spatial relations or generic places; they explain why things occur where they do (Sack 1974). Thus the generalizations from other disciplines would adequately address the questions raised by generic place. This is why, in anticipation of the spatial analysis position of Schaefer (1953) and Bunge (1966) that urges geography to concentrate on laws about spatial relations, Hartshorne (1939, 127, 417–18) argues that geography would not be distinct if it simply wanted to explain the "where" of things or their distributions. Any science could do that.

From the nomothetic perspective, the specific place can be thought of as residuals or as manifestations of contingent relations in open systems. These raise problems for nomothetic science, which examines fundamental forces and expresses them in theories or models. Of course, there are other forms of explanation that imply causal links, such as explanation based on reasons, but these and other processes of objectification still require that appeals be made to more general contexts (Nagel 1986).

Hartshorne's course, then, is to skirt the general and have chorology be a method of analysis suited to the specific. He argues that we can use generalizations to initiate our investigations, but we must not conclude that generalizations address all of the important relationships under study. What then do we need to use instead? Here Hartshorne offers suggestions about the criteria of significance that chorology might employ. He argues (1939, 237–49) that chorology's criteria are not the nomothetic notion of empirical association, but rather involve components such as "significant to areal differentiation" and "significant to man."

These criteria are very much like the ones that historians claim to use in their analyses. This is why geography and history seem to have so much in common. Just as the geographer studies particular places, the historian studies particular eras and epochs. And just as places are composed of processes, so too are these periods of time. The process constituting the historical period could be examined under the light of generalizations, but because the particular processes of these periods are so complex and special, the generalizations would not do them justice. Thus the French and American wars of independence can be analyzed as instances of revolutions or even of bourgeois revolutions. Seeing them as such is important, but inadequate to express their true characters.

The criteria for chorology that Hartshorne advocates, of significance to areal differentiation and significance to man, do differ from the nomothetic criteria which rely on empirical associations. The nomothetic sciences prize generalizations and the power of explanation and prediction. The chorological perspective may do so also, but, again like historical analysis, it values equally, or perhaps even more, the unusual and the different. In this respect, our own intent and values as observers become basic issues in the construction of specific places. Hartshorne's criteria of significance show that he believes the analysis of specificity to be as much a product of the mind as a product of fact. The chorological perspective is used to disclose the specific qualities of things because these are valued (remember that for Hartshorne geography is there to satisfy a natural curiosity about specific places), and such qualities would disappear without this perspective. Thus the mind, through the high value it places on the specific, becomes an active agent in its construction. Much of the specific place, then, results from a point of view (1939, 425).

## Reactions to *The Nature*

These, then, are the strands from the levels of perspective and forces that Hartshorne has woven together. We much now consider how his arguments appear in light of subsequent geographic research in these areas. Most of the recent developments would loosen and broaden Hartshorne's synthesis, whereas the spatial-analytic tradition would tighten and narrow it. Let us consider first the spatial analysis position and then survey the more recent research.

The primary device used to tighten the argument is to replace Hartshorne's coexistence of the ideographic and nomothetic meanings of science with just the narrower and more distant view from nowhere—the nomothetic (Schaefer 1953; Bunge 1966). Emphasis on the nomothetic focuses attention on more restricted methods of verification and (as Hartshorne anticipated in his arguments against defining geography as the study of location, esp. 127–30) leads to a narrower meaning of place—the generic. Geography then becomes the nomothetic explanation of the where of things or of their spatial relations. This position has distinct benefits. It would place geography among the other nomothetic sciences. Geography would acquire more prestige because our society invests the discovery of laws and theories with enormous status. Even if the search for laws fails, the quest presents a single unified perspective from which to view geography and also a single set of mostly quantitative methodologies. Method, if nothing else, can provide geographers with common ground. But it also has costs, as Hartshorne anticipated, in that it loses geography's identity. Careful analysis of testable laws and theories from any discipline shows that they are all spatial in their content, and thus geographical.

Most recent developments in geography have broadened the scope of perspectives from nowhere back to somewhere and of the variety of forces in each realm; these have made Hartshorne's synthesis appear relatively narrow. Consider first the changes within the level of perspectives. The mode of analysis that geographers use is still objective and discursive, but, where once the range was limited to the nomothetic and the ideographic, it now includes dialectics and positions like empathy, existentialism, phenomenology, and narrative (Tuan 1971, 1972; Entrikin 1976; Gregory 1978; Ley and Samuels 1978; Guelke 1982) that are closer to somewhere. Not all of these are objective in the same way, but they all attempt to see the world from a perspective others can share. Each brings particular facets of space and place into view, and a theoretical task is to understand how the perspectives and the qualities of space and place are interrelated. The perspectives have even been used to view each other, as when geographers examine how the phenomenological perspective is seen from the nomothetic view (Entrikin 1976, forthcoming 1990). Moreover, the perspectives have been used to examine the structure of other views, such as art, magic and ritual, and the child's view, that are not even employed by professional geographers but are used in everyday life to organize and give meaning to space (Sack 1980). These views have changed historically in character, relative importance, and interconnections. One of the major tasks in geography is to construct a general history of these perspectives and their connections to the changes in the particular realms. There is no doubt that perspectives and forces are historically interdependent.

Perspectives can themselves be forces, as Hartshorne well knows, and geographers are considering far wider ranges of elements within each realm of force. Geographers are now interested in how landscapes can be constituted by mythical, magical, and religious beliefs, and how they can be built according to aesthetic impulses. The modern world, though, relies most on the nomothetic form of science to plan and construct place. The nomothetic perspective can transform the landscape in certain directions. The stronger this emphasis becomes, the more it pushes aside interest in the specific place and focuses attention on a geometric continuum.

An equivalent broadening and deepening has occurred in the realm of social relations. Hartshorne acknowledges that states and economies play an important role in affecting the element complexes, but geographers now consider a range of social theories attempting to identify the key elements in the political economy of a period that drive other elements from other realms and even the perspectives of the times. Perhaps the most ambitious of these is Marxist theory, which draws attention to class conflict and the mode of production as the locus of energy that, in the last instance, affects the realm of meaning, creates a "second nature" (Smith 1984), and creates places as territories that are useful to capital (Harvey 1982, 373; Sack 1986, 87–91). Other social theories have revealed the power of particular elements in the political economy to shape our ideas and to alter nature. Indeed, there seems to be a consensus that the development of a global economy and culture has been a force making the world more homogeneous and places more generic. In this respect, both modern social forces and the force of the nomothetic perspective in science lead to similar and mutually reinforcing results, that is, the homogenization of space and the marginalization of the specific place (Entrikin forthcoming 1990).

Similar increases in scope have occurred within the realm of nature. In place of the notion of direct one-way causal links between nature and behavior in theories of environmental determinism (or probabilism or possibilism) are the more subtle and complex feedback loops of modern human ecology (Ellen 1982). These permit a deeper and more precise description of the penetration of natural forces into the realm of meaning and social relations.

As far as the role of agents is concerned, nature, meaning, and social relations contain agents as instruments of their powers, but none of these perspectives is very comfortable with the idea of having agents whose behavior is, in some measure, a matter of free will. Indeed, as we have argued, it may be impossible to retain a sense of free will once actions are examined from a perspective that is outside ourselves. Reasons, causes, and forces in general reduce free will to an epiphenomenon.

Hartshorne recognizes, but does not stress, the role of individual actors in creating places and considers free will, but in terms of how it thwarts nomothetic explanations. Some have given agency greater attention, although they have not confronted the issue of free will directly. Humanistic geographers focus primarily on intention and meaning in the creation of place, and structuration theory attempts to avoid several oppositions by conceiving of structure and agency as mutually constitutive. Here though, as we noted, it seems that agents, who are simultaneously constrained and enabled by structures, are not exercising will but are being propelled by forces that are either contingent or that have not been specified at a particular level.

An emphasis on agents, even as intruments of forces if not as free individuals, makes vivid the fact that places are created and sustained by both the extraordinary and the ordinary, everyday, labors of human beings (Gregory 1985). Examining individual actions in great detail has reignited an interest in non-nomothetic forms of explanation (such as singular causation, Entrikin forthcoming) and also an interest in the specific place (Agnew 1987, 42; Massey and Allen 1984, 299–300; Sayer 1984, 115–16). The connections among agent, non-nomothetic explanations, and specific places repeats Hartshorne's attempts to use method as both a force and a perspective. Shunning generalizations leaves us with a view of human nature as open and contingent. As with Hartshorne's chorological view, this indefiniteness about the significant forces operating in a particular context leads to a focus on the details of human actions that make the entire context in which actions occur seem specific or unique. But exactly how these particular actions are connected and form places and landscapes has not been worked out and thus far requires ad hoc assumptions about the significance of proximity (or distance) and assumptions concerning structure and scale (e.g., locality space/time, Giddens 1984, 118–22; Pred 1986, 6) that make sense only if we appeal to sketches of generalizations about the relationship between space, or place, and particular kinds of actions. In other words, space enters into the picture through unexamined generalizations about its effect.

A focus on agency also draws attention to the fact that not only are places constantly sustained and transformed through individual actions, but that space or place appear to help constitute and transform these very actions and forces.

While this dualism seems to make sense, geography has been able more clearly to show how geographical space, place, and landscape are products of forces than how they themselves become forces. How space and place can become forces is now emerging as among the most important theoretical problems in geography. If there are four realms of force, would space and place be involved in each of them, and would it be the same kind of space and place? Whereas the field has not yet developed a systematic approach to these questions, at this point it does appear that geographers have been thinking that the answers lie in the affirmative.

In the realm of nature, analysis reveals that physical geographic space cannot have an independent effect except insofar as it is related to particular substances or media through which energy can be transmitted. This is the "relational" concept of space (Sack 1980, 59–85) and shows how space is inextricably intertwined with substance. Such an embedding of space with substance does not make space an independent force, but rather an equal partner in constituting force. The relational concept is an example within the realm of nature that also applies to the other realms.

As for the realm of meaning, there already exists a strong tradition in neo-Kantianism of how space, as a synthetic a priori, becomes inextricably intertwined with thought. This tradition moreover is useful not only for mental powers but also for perspectives, and allows for a variety of meanings of space (see Cassirer 1953; Langer 1953; Sack 1980) that need not be idealist. Neo-Kantian approaches might not be the only device for understanding the power of space in the realm of meaning, but thus far they have borne rich geographic fruit.

Things are less clear with space as a force in the realms of social relations and agency. Because few have directed attention to the question of free will, and because agents, whether free or not, must be the vehicles for all social actions, I will pass over space in agency (except to note that humanistic geography makes "being" and "place" practically synonymous) and focus attention on the power of space in social relations. One approach to having space become an agent in social relations is found in the spatial analytic tradition which attempts to make physical geographic space, through distance, a direct variable or force in human action. But careful analysis demonstrates that, as in the realm of nature, geographical space in social relations must comply with the relational concept. When it does, it becomes an equal partner with substance in constituting a force. The relational concept then applies both to the natural and to the social realm. Some have called for a more distinct conception of the power of space in social relations—a "spatiality" (Soja 1989)—that "can be distinguished from the physical space of material nature and the mental space of cognition and representation" (Soja 1989, 120). While these other realms contribute to the character of spatiality, spatiality's advocates, who also give primacy to the realm of social relations, see spatiality as having considerable effect over these other realms. The real question is what distinguishes spatiality and how is it embedded in social relations? The answer to this falls far short of ontology and focuses on a narrower issue, the necessity of spatial inequalities in the social relations of capitalism (Soja 1989, 107, 113). Not only is the argument narrower than the question of ontology, it also does not prove necessity. For example, it is conceded that, as a special case, capitalism can exist without significant areal variation of inequality (Soja 1989, 113) and, more importantly, the explanation of the power of unevenness presupposes the effects of agglomeration and distance decay wherein space works relationally. In this way, the relational concept allows physical space to enter into social relations. Although this explanation does not seem to be what the spatiality advocate means by the power of space, spatiality itself offers no other alternatives. Rather it becomes yet another way of saying, but not showing, that our actions are both constitutive of space and constituted by space.

What then could be the basic role of space in social relations? Does it have to be different from a physical-relational space and a neo-Kantian space of the mind? And how could these all be interrelated? I believe these are basic questions confronting geography, and I offer no solutions here except to point out that yet another assumption that spatiality makes can help point us to spatial concepts that might just integrate these realms. I am referring here to the correct, yet unexamined, assumption of spatiality that the world is divided into territorial units of varying sizes and types. Hartshorne, in *The Nature*, addresses a special type of territoriality, the political state (1939, 401–05 esp.), but a more general conception of territoriality shows it to be the primary way in which space is constructed and by which it helps exert power. Territoriality is the "attempt by an individual or group to affect, influence, or control people, phenomena, and relationships, by delimiting and asserting control over a geographic area" (Sack 1986, 19). Territoriality then is a spatial strategy to exert power, and obviously it is embedded within the actions of individuals or agents. It also draws on their values and attitudes, that is, their realm of meaning. And territoriality involves the physical structure of space and its relational effect on interaction. But because territoriality is an assertion of control, it lends power to this constructed space.

Territoriality is but one unexamined spatial factor in most theories of social relations. Other spatial concepts could be developed that would show how the realms are interrelated. These would help us clarify not only Hartshorne's questions about the way in which place integrates the realms and perspectives, but also the way in which place is a force affecting these realms and perspectives.

## Conclusion

*The Nature* provides a tightly woven and confident but inward-looking analysis about geographic reality and method. In the light of the present, it seems that much of *The Nature*'s precision and inward quality comes from the many factors that it held "constant" but which now have been allowed to vary all together. The expansion of perspectives from somewhere to nowhere, and of the realms of meaning, nature, social relations, and agency, result in a dizzying array of alternatives that is characteristic not only of contemporary geography, but also of the contemporary intellectual world. This world in a sense is losing its center. An extremely ambitious challenge to geography is to map out the ranges and interconnections among the forces and perspectives, and to explain the indispensability of space and place to them all. If place can be shown to be the locus for all of them, then fifty years after *The Nature*, geography will again have a clear focus that is central to contemporary intellectual concerns.

# References

**Agnew, J.** 1987. *The politics of place.* London: Allen Unwin.

**Bunge, W.** 1966. *Theoretical geography.* Lund Studies in Geography, Series C. General and Mathematical Geography, No. 1. Lund: C. W. K. Gleerup.

**Cassirer, E.** 1953. *Philosophy of symbolic forms,* 3 vols. New Haven, CT: Yale University Press.

**Cosgrove, D.** 1984. *Social formation and symbolic landscape.* London: Croom Helm.

———, **and Daniels, S., eds.** 1988. *The iconography of landscape.* Cambridge: Cambridge University Press.

**Daniels, S.** (Forthcoming). Marxism, culture and the duplicity of landscape. In *New models in geography,* ed. R. Peet and N. Thrift. London: Methuen.

**Ellen, R.** 1982. *Environment, subsistence and system.* New York: Cambridge University Press.

**Entrikin, J. N.** 1976. Contemporary humanism in geography. *Annals of the Association of American Geographers* 66:615–32.

———. (Forthcoming). Place, region and modernity. In *The power of place,* ed. J. Agnew and J. Duncan. Boston: Unwin Hyman.

———. (Forthcoming, 1990). *The betweenness of place.* Baltimore: Johns Hopkins University Press.

**Freud, S.** 1952. *Civilization and its discontents.* Chicago: Great Books.

**Giddens, A.** 1984. *The constitution of society.* Berkeley: University of California Press.

**Gregory, D.** 1978. *Ideology, science and human geography.* London: Hutchinson.

———. 1985. *Space and time in social life.* Worcester, MA: Clark University Graduate School of Geography, Wallace W. Atwood Lecture Series No. 1.

**Guelke, L.** 1982. *Historical understanding in geography.* Cambridge: Cambridge University Press.

**Hartshorne, R.** 1939. *The nature of geography: A critical survey of current thought in the light of the past.* Lancaster, PA: Association of American Geographers (reprinted with corrections, 1961).

———. 1959. *Perspective on the nature of geography.* London: Rand McNally.

**Harvey, D.** 1982. *The limits to capital.* Oxford: Basil Blackwell.

———. 1985. *The urbanization of capital: Studies in the history and theory of capitalist urbanization.* Baltimore: Johns Hopkins University Press.

**Jackson, P.** 1988. Social geography; social struggles and spatial strategies. *Progress in Human Geography* 263–69.

**Langer, S.** 1953. *Feeling and form.* New York: Charles Scribner's.

**Lévi-Strauss, C.** 1966. *The savage mind.* Chicago: University of Chicago Press.

**Ley, D., and Samuels, M., eds.** 1978. *Humanistic geography: Prospects and problems.* Chicago: Maaroufa Press.

**Massey, D., and Allen, J.** 1984. *Spatial divisions of labor: Social structures and the geography of production.* New York: Methuen.

**Nagel, T.** 1986. *The view from nowhere.* New York: Oxford University Press.

———. 1987. *What does it all mean?* Oxford: Oxford University Press.

**Piaget, Jean, and Inhelder, Barbel.** 1956. *The child's conception of space.* New York: W. W. Norton.

**Pred, A.** 1986. *Place, practice and structure.* Totowa, NJ: Barnes and Noble.

**Sack, R.** 1974. Chorology and spatial analysis. *Annals of the Association of American Geographers* 64:439–52.

———. 1980. *Conceptions of space in social thought: A geographic perspective.* Minneapolis: University of Minnesota Press.

———. 1986. *Human territoriality: Its theory and history.* Cambridge: Cambridge University Press.

———. 1988. Place as context: The consumer's world. *Annals of the Association of American Geographers* 78:642–64.

**Sayer, A.** 1984. *Method in social science: A realist approach.* London: Hutchinson.

**Schaefer, F.** 1953. Exceptionalism in Geography. *Annals of the Association of American Geographers* 43:226–49.

**Smith, N.** 1984. *Uneven development: Nature, capital, and the production of space.* Oxford: Basil Blackwell.

**Soja, E.** 1989. *Postmodern geographies: The reassertion of space in critical social theory.* London: Verso.

**Thrift, N.** ms. Images of social change.

**Tuan, Yi-Fu.** 1971. Geography, phenomenology and the study of human nature. *The Canadian Geographer* 15:181–92.

———. 1972. Structuralism, existentialism, and environmental perception. *Environment and Behavior* 4:319–31.

———. 1977. *Space and place: The perspective of experience.* Minneapolis: University of Minnesota Press.

———. 1989. Surface phenomena and aesthetic experience. *Annals of the Association of American Geographers* 79:233–41.

# Epilogue: Homage to Richard Hartshorne[1]

## DAVID R. STODDART

Department of Geography, University of California at Berkeley,
Berkeley, CA 94720

It may perhaps seem paradoxical that, half a century on, I should write from Carl Sauer's old department at Berkeley, beyond the High Sierra, to pay homage to Richard Hartshorne in celebration of the publication of *The Nature of Geography*. Hartshorne was in fairly continuous disagreement on intellectual matters, not only with Sauer but also with his colleague John Leighly, from the middle 1930s to the end of their lives. In many respects the Berkeley tradition was unsympathetic, if not antithetic, to Hartshorne's endeavor. Needless to say, I write in no spirit of atonement for those disagreements. But I do believe that *The Nature of Geography* is one of the most remarkable contributions to geographical scholarship in this century, and Hartshorne himself a scholar of stature and authority. There are four components in this belief.

First, Hartshorne with this book brought intellectual maturity to American geography. Certainly no one had previously attempted such a sustained and penetrating work of scholarship before, and indeed few since. Hartshorne (1979) himself has described how the book came to be written: it is daunting to think that this monograph of 480 pages was produced from inception to publication in only some twenty months, during much of which the author was constantly traveling in an increasingly embattled Central Europe. That a work of such length could be published in the *Annals* at all says much for the editorial audacity of Derwent Whittlesey, who indeed suggested that Hartshorne write it in the first place. It is difficult to think of any Council permitting such expenditure these days. It was indeed, as Whittlesey admitted, an "unprecedented step," justified by the fact that he found the work "both timely and timeless" (Whittlesey 1939, 171–72). In Thomas Kuhn's later terminology, *The Nature of Geography* defined a paradigmatic position, validated historically but also claiming programmatic significance. And as Kuhn's analysis would also suggest, after dominating methodological discussion for years, *The Nature* almost inevitably provoked its own reaction.

The second point is that Professor Hartshorne virtually single-handedly internationalized American geography. Indeed, for at least three decades (and perhaps still), the majority of geographers knew what they did of Humboldt, Ritter, Richthofen, Hettner, Gerland and the rest from the pages of *The Nature of Geography* and its ancillary papers (notably Hartshorne 1958), either directly from Hartshorne's own work or from writings derivative from it. In the subsequent *Perspective on the Nature of Geography* (1959), Hartshorne also allowed that there was something to be learned from the French, English and Scottish geographers too, and there will likewise be many who owe their knowledge of this work to Hartshorne's writings. Few could match either his scholarship or his command of the German language, with the possible exception of Leighly. But while two of Leighly's earlier papers (1937, 1938) had provided the immediate stimulus for *The Nature,* he retired from this particular fray, at least in public, after Hartshorne's book was published.[2] But certainly, after the appearance of *The Nature,* no American scholar could subsequently be ignorant of the European underpinnings and ancestry of the subject.

Third, *The Nature* had a remarkable effect overseas: in a sense, it stimulated the emerging coherence of world geography as an intellectual discipline, at least in the English-speaking realms. It was the first time that a recognizably modern American geography had made an impact on the heartlands of traditional geography in Europe, even though constrained by the Second World War. Henceforth the terms of the debate were those defined by Hartshorne. This became clear as soon as the war was over, most notably in Wooldridge's inaugural lecture at Birkbeck College, London, in 1945. Wooldridge followed Hartshorne closely, quoted his translation of Hettner's view of the place of geography among the sciences, and gave his opinion that, after Hartshorne, the definition of the aims of the subject was no longer in question (Wooldridge 1956, 9–10, 20). I think it would be true to say that, certainly in Britain for the next fifteen or twenty years, every single inaugural lecture by a new chair of department was based upon the premises that Hartshorne had defined in *The Nature of Geography.* Such was the impact of what a contemporary called "this leviathan of learning."

And finally, of course, there was the reaction to both the thesis and the method of *The Nature,* already well under way by the time that *Perspective* was published in 1959. In interpreting the nature of our subject, all of us during our professional lives have been concerned by questions of philosophy, history, epistemology, contextuality, and their mutual and reciprocal relations. Many of these questions, arising long after *The Nature* was published, were undoubtedly provoked by it: indeed *The Nature* itself provides their own contextuality.

It is all too easy, with the hindsight of history, to apply T. H. Huxley's model of the reception of innovations to such a case as Hartshorne's. For Huxley, innovators shortly come to be seen as people of sublime genius and perfect virtue. But then the euphoria passes: the innovation will not explain things in general after all and is therefore a wretched failure. But in the longer term, it is seen to explain as much as could reasonably be expected, and its propounder to be a person worthy of all honor in spite of his share of human frailties, as one who has added to the permanent possessions of science (Stoddart 1986). This is a caricature, of course, but it would be of great interest to look as closely at the reception of *The Nature* as Hartshorne himself has looked at its origins.

What strikes me throughout is the generative power of major works of scholarship such as this. *The Nature of Geography* represented for the first time in North America a new, meticulous, and rigorous standard which had not been seen before. The standards (and indeed, I would like to say, the quiet good manners of true scholarship) were shown to perfection both in this work and in *Perspective*. Hartshorne showed himself acutely alive to the issues involved (Hartshorne 1948).The assignment of credit where credit is due, through the technical canons of scholarship, but also the assignment of responsibility where responsibility lies, have been hallmarks of Hartshorne's work, not least in his repeated acknowledgments to Sauer. Always he has sought in his published work to avoid personal rancor and the trivialization of issues which this implies. This is not to say that Hartshorne has not been tenacious in defense of his views—for some, overly so. But he has always acknowledged his intellectual indebtedness, perhaps especially to those with least sympathy for his arguments and least patience with his methods. In his whole scholarly output throughout his career, he has given American geography a new sense of scholarly obligation: to know where we stand, why we are there, where we are going, coupled with a necessity to be able to explain and justify our positions to our professional peers.

Thus, in reaching a conclusion about the contribution of *The Nature of Geography*, I would like to make a distinction between the analysis and critique of its substantive content and its argument and the function of the book in the evolution of our subject. Here is a book which continues to live, even though today we approach it on terms other than its own. The reason we celebrate the passage of fifty years since its publication is not merely to mark an anniversary. It is because Dick Hartshorne marked out for us some of the primary questions. He suggested ways in which they might be answered. Perhaps the ways of answering them have changed; the questions, of course, remain. Dick Hartshorne knows well enough that geography has moved on in the last fifty years: indeed it would be appalling if this were not so. Many of us doubtless feel that he charted a trail that we do not now recognize as we look back over the course of each of our intellectual traverses. But in insisting on rigorous, exact, and supportable explanations of what it is that we are about, he supplied a benchmark of scholarly integrity from which the subject has advanced.

Dick Hartshorne: a man of standards, therefore, a man, in his time, of great originality. The standards he set were for himself but also for us. He has touched profoundly the evolution of our discipline and our understanding of it. Fifty years after the appearance of *The Nature of Geography*, he is to be saluted for it.

## Notes

1. Based on informal remarks at the conclusion of the session marking the 50th anniversary of the publication of *The Nature of Geography*, AAG Annual Meeting, Baltimore, March 20, 1989. I am very grateful to Maynard Weston Dow, Plymouth State College, for providing me with a transcription of these remarks from his series *Geographers on Film*, and to David Ward, AAG President, for the invitation to make them on that occasion.

2. Leighly, however, embarked on an extensive critique of *The Nature* in a vigorous correspondence with Hartshorne through November and December 1939.

## References

**Hartshorne, R.** 1939. The nature of geography. A critical survey of current thought in the light of the past. *Annals of the Association of American Geographers* 29:173–658, issued as a single volume, 1939, reprinted with corrections 1961. Lancaster, PA: AAG.

——. 1948. On the mores of methodological discussion in American geography. *Annals of the Association of American Geographers* 38:113–25.

——. 1958. The concept of geography as a science of space, from Kant and Humboldt to Hettner. *Annals of the Association of American Geographers* 48:97–108.

——. 1959. *Perspective on the Nature of Geography.* Chicago: Rand McNally.

——. 1979. Notes toward a bibliobiography of *The Nature of Geography. Annals of the Association of American Geographers* 69:63–76.

**Leighly, J.** 1937. Some comments on contemporary geographic method. *Annals of the Association of American Geographers* 27:125–41.

——. 1938. Methodologic controversy in nineteenth century German geography. *Annals of the Association of American Geographers* 28:238–58.

**Stoddart, D. R.** 1986. *On Geography and its history.* Oxford: Blackwell.

**Whittlesey, D.** 1939. A foreword by the editor [to *The Nature of Geography*]. *Annals of the Association of American Geographers* 29:171–72.

**Wooldridge, S. W.** 1945. *The geographer as scientist.* London: University of London. Reprinted in *The geographer as scientist: Essays on the scope and nature of geography.* London: Nelson.

# Index of Proper Names